THE ORIGIN AND EVOLUTION OF THE UNIVERSE

EDITED BY

BEN ZUCKERMAN
MATTHEW A. MALKAN

JONES AND BARTLETT PUBLISHERS
Sudbury, Massachusetts

Boston   London   Singapore

Editorial, Sales, and Customer Service Offices

Jones and Bartlett Publishers
40 Tall Pine Drive
Sudbury, MA 01776
(508) 443-5000
(800) 832-0034

Jones and Bartlett Publishers International
Barb House, Barb Mews
London W6 7PA
UK

Copyright © 1996 by Jones and Bartlett Publishers, Inc.

All rights reserved. No part of the material protected by this copyright notice may be reproduced or utilized in any form, electronic or mechanical, including photocopying, recording, or by any information storage and retrieval system, without written permission from the copyright owner.

Library of Congress Cataloging-in-Publication Data
The origin and evolution of the universe / edited by Ben Zuckerman, Matthew A. Malkan.
     p.   cm.
    Includes bibliographical references and index.
    ISBN 0-7637-0030-4
    1. Cosmology. I. Zuckerman, Ben, 1943-   . II. Malkan, Matthew Arnold.
QB981.O72   1996
523.1--dc20
                                     96-4260
                                        CIP

Acquisitions Editor: David E. Phanco
Associate Production Editor: Nadine Fitzwilliam
Senior Manufacturing Buyer: Dana L. Cerrito
Editorial Production Service: Ocean Publication Services
Typesetting: Ruth Maassen
Cover Design: Hannus Design Associates
Printing and Binding: Courier Companies Inc.
Cover Printing: John P. Pow Company, Inc.

A Contribution of the IGPP Center for the Study of Evolution and the Origin of Life (CSEOL), University of California, Los Angeles

Printed in the United States of America
00 99 98 97 96   10 9 8 7 6 5 4 3 2 1

CONTENTS

CONTENTS   v

PREFACE   xi

BIOGRAPHICAL SKETCHES   xiii

CHAPTER 1   THE ORIGIN OF THE UNIVERSE   1
Edward L. Wright

INTRODUCTION   1

EXPANSION OF THE UNIVERSE   1

COSMIC BACKGROUND   5

LIGHT ELEMENT ABUNDANCES   7

HORIZONS   8

INFLATION   9

CURRENT RESEARCH   11
   Dark Matter   11
   Age and the Hubble Constant   12
   Infrared Background   12

CONCLUSION   13

FURTHER READING   13

REFERENCES   13

CHAPTER 2   THE ORIGIN AND EVOLUTION OF GALAXIES   15
Alan M. Dressler

INTRODUCTION   15

WHAT IS A GALAXY?   16

Contents

HOW DID GALAXIES FORM? — THE SHORT VERSION   20

MAKING REAL CALCULATIONS OF HOW GALAXIES FORM   23

USING THE HUBBLE SPACE TELESCOPE AND N-BODY COMPUTERS TO GO FURTHER   26

COMPLEXITY — UNDERSTANDING THE EVOLUTION OF THE UNIVERSE AND TURNING THE COPERNICAN REVOLUTION   28

FURTHER READING   35

CHAPTER 3   THE ORIGIN OF STARS AND PLANETS   37

Fred C. Adams

INTRODUCTION   37

STAR FORMATION IN MOLECULAR CLOUDS   38

    Molecular Clouds — The Birthplace of Stars   39

    Core Formation Through Ambipolar Diffusion   40

    Molecular Cloud Cores — Initial Conditions for Collapse   42

    Protostellar Collapse   42

    Protostellar Radiation   43

    The L–A_V Diagram for Protostars   44

    The Protostellar to Stellar Transition   46

    Pre-Main-Sequence Stars and Circumstellar Disks   47

    Summary of the Star-Formation Paradigm   49

PHYSICS OF CIRCUMSTELLAR DISKS   49

    Gravitational Instabilities   50

    Nonlinear Simulations of Star/Disk Systems   51

    Viscous Evolution of Circumstellar Disks   53

    Additional Disk Processes   53

    Summary of Disk Processes   54

PLANET FORMATION   54

    Formation of Planets by Accumulation   55

    Formation of Planets by Gravitational Instability   56

THE INITIAL MASS FUNCTION   57

    The IMF Observed   57

    The IMF as a Two-Phase Process   58

    A Scaling Relation for the IMF   59

SUMMARY AND DISCUSSION   60

REFERENCES   61

CHAPTER 4   **STELLAR EXPLOSIONS, NEUTRON STARS, AND BLACK HOLES   65**

Alexei V. Filippenko

INTRODUCTION   65

STELLAR EXPLOSIONS — CELESTIAL FIREWORKS   67

HOW TO FIND A SUPERNOVA   68

SUPERNOVA CLASSIFICATION   69

    Explosion Mechanism — Type I   71

    Explosion Mechanism — Type II   72

SUPERNOVA 1987A — A GIFT FROM THE HEAVENS   73

    Testing the Theories   74

    Neutrinos from Hell   75

NEUTRON STARS   77

    Observational Evidence   79

    Millisecond and Binary Pulsars   81

BLACK HOLES   81

    Fun Facts about Black Holes   83

    Detecting Black Holes   84

    Myths about Black Holes   85

CONCLUSION   86

FURTHER READING   87

CHAPTER 5   **THE ORIGIN AND EVOLUTION OF THE CHEMICAL ELEMENTS   89**

Virginia L. Trimble

INTRODUCTION   89

BIG BANG NUCLEOSYNTHESIS   90

ELEMENT SYNTHESIS IN STARS   93

 The Major Burning Phases — Hydrogen   94

 The Major Burning Phases — Helium   96

 The Major Burning Phases — Heavy Elements   97

 On Beyond Iron   99

 A Few Loose Ends   101

GALACTIC CHEMICAL EVOLUTION   102

 Simple Models and the G Dwarf Problem   103

 Models with Bells and Whistles   104

 Stellar Populations and the Incidence of Habitable Planets   106

REFERENCES   107

CHAPTER 6   THE ORIGIN AND EVOLUTION OF LIFE IN THE UNIVERSE   109

Christopher P. McKay

INTRODUCTION   109

LIFE ON EARTH   109

LIFE ON MARS   117

ANTARCTICA AND MARS   119

LIFE IN THE SOLAR SYSTEM   120

SEARCHING FOR LIFE AROUND OTHER STARS   122

IT'S LIFE, JIM, BUT NOT AS WE KNOW IT   123

REFERENCES   124

CHAPTER 7   FUTURE OF THE UNIVERSE   127

Andrei Linde

BIG BANG THEORY VERSUS INFLATIONARY COSMOLOGY   127

BIG BANG THEORY AND THE FUTURE OF THE UNIVERSE   128

PROBLEMS OF THE BIG BANG THEORY   129

INFLATIONARY COSMOLOGY   130

 Unified Theories of Elementary Particles   130

Contents

The Simplest Version of Inflationary Theory
— Chaotic Inflation   132

Brief History of Inflationary Cosmology   133

Quantum Fluctuations as the Origin of Structure Formation
in the Universe   134

Experimental Tests of Inflationary Theory   135

Self-Reproducing Inflationary Universe   136

BACK TO THE FUTURE   138

REFERENCES   139

GLOSSARY   141

INDEX   147

PREFACE

Astronomy is in the midst of its second golden age. The first golden age was in the 1600s when Galileo and other astronomers used the optical telescopes that had just been invented. Now, the second golden age is a result of the development of radio, infrared, X-ray, ultraviolet, gamma-ray, and cosmic ray astronomy, all since World War II.

Thanks to these new technologies, the rate of discovery in astronomy is at an all-time high. Galileo also had a new technology, his optical telescope. But there is at least one very significant difference between then and now. Astronomers of the 1600s did not have to rely on federal and state funding — on society's tax dollars — to support their research. In that sense, virtually all the wondrous discoveries described in this book can be credited in large part to citizens, not simply to the astronomer who happened to be at the telescope.

On March 17, 1995, a symposium, "The Origin and Evolution of the Universe," was held on the campus of the University of California, Los Angeles. The symposium, convened by UCLA's Center for the Study of Evolution and the Origin of Life, was attended by an all-day overflow audience of diverse elements of the Los Angeles community including college students, faculty, researchers and other members of the UCLA family, and members of the public. The topics of the seven presentations, which are included in this volume, ranged from the very large — the Universe and beyond — to the very small — life; from the very young — the birth of stars that is occurring at this moment — to the very old — the birth of the Universe more than 10,000 million years ago; from the here and now to distant times and places.

Some of the fundamental questions have not changed greatly over the centuries — we are still seeking to understand our place in the cosmos — but the answers we can now give are far deeper and richer than those offered previously. Only a few decades ago, the idea of a book on the origin and evolution of the Universe would have seemed outrageously audacious. Today, it is still an ambitious task because it necessarily spans several scientific disciplines, from quantum physics and relativity to chemistry, geology, and biology. Fortunately, we were able to bring together researchers with the broad range of expertise needed to do justice to such a grand topic. These researchers are active participants in forefront areas of astronomy and are well qualified to present the latest insights on the basis of discoveries that, oftentimes, have been made within the past few years. All the authors take very seriously their duty to convey these exciting developments to the broader public.

The book begins with a discussion of the origin of the Universe by Dr. Ned Wright, followed by a description of the origin and evolution of the galaxies by Dr. Alan Dressler. Stars are a major constituent of galaxies; Dr. Fred Adams analyzes the formation of stars and planetary systems. Stars live and die, often in spectacular fashion, and this evolution is considered by Dr. Alex Filippenko. Accompanying all this activity is the synthesis of chemical elements essential for building planets and living creatures — the origin and evolution of the elements is discussed by Dr. Virginia Trimble. Next,

Dr. Chris McKay ponders the question of life in our solar system and elsewhere in the Universe. Finally, Dr. Andrei Linde speculates on the origin and future of the Universe (or universes) in light of current ideas in elementary particle physics and cosmology.

This book is aimed to satisfy readers with a wide range of preparations and interests. It presents up-to-date discussions of cutting-edge advances in astronomy and physics that will intrigue students and amateur and professional practitioners in all the sciences. Essential physical concepts are explained at approximately the knowledge level of a college freshman. Mathematical detail has been deliberately kept simple to make the chapters accessible to anyone with an appreciation of science, including students and former students not majoring in the sciences.

BIOGRAPHICAL SKETCHES

Chapter 1: The Origin of the Universe
EDWARD L. WRIGHT

Edward (Ned) Wright received his A.B. *summa cum laude* in physics from Harvard College, then spent one year at the Naval Research Laboratory working on the long-range propagation of underwater sound before he began graduate studies in astronomy at the Harvard College Observatory. He was a Junior Fellow in the Harvard Society of Fellows while finishing his dissertation, which was based on observations made with a 102-cm-diameter telescope that was carried to an altitude of 30 km by a helium-filled balloon. After receiving his Ph.D. in astronomy, he was an Assistant and Associate Professor of Physics at MIT, and he came to UCLA as Professor of Astronomy in 1981. Since 1978, Dr. Wright has been a member of the Science Working Group for the COsmic Background Explorer (COBE), an Earth-orbiting satellite, which measured the radiation produced by the Big Bang. He has been an Associate Editor of *The Astrophysical Journal* since 1994.

Chapter 2: The Origin and Evolution of Galaxies
ALAN M. DRESSLER

Alan Dressler is an astronomer at the Observatories of the Carnegie Institution in Pasadena, California. He received his Ph.D. degree in astronomy from the University of California, Santa Cruz, in 1976. Dr. Dressler's research focuses on the structure and composition of galaxies and how these have changed through cosmic time. One of the first to find observational evidence for massive black holes in galactic centers, he was also a principal contributor to the study that led to discovery of the "Great Attractor," a supercluster-sized mass concentration that perturbs the nature of space, a study now extended to an attempt to map the distribution of the "dark matter" that dominates the Universe. Recipient of the Newton Lacy Pierce Prize of the American Astronomical Society, he is author of *Voyage to the Great Attractor* (1994), the story of the search for the Great Attractor and other recent advances in human understanding of the Universe.

Chapter 3: The Origin of Stars and Planets
FRED C. ADAMS

A theoretical astrophysicist, Fred Adams received his undergraduate training in mathematics and physics at Iowa State University and, in 1988, his Ph.D. in physics from the University of California, Berkeley. After serving as a Research Fellow at the Harvard-Smithsonian Center for Astrophysics, he joined the faculty of the Physics Department at the University of Michigan in 1991. Recipient of both a National Science Foundation Young Investigator Award and the Robert J. Trumpler Award of the Astronomical Society of the Pacific, his research focuses on star formation and cosmology. Dr. Adams is internationally recognized for his studies of the radiative signature of the star formation process, the dynamics of circumstellar disks, and the physics of molecular clouds. In cosmology, he has made contributions to understanding of the inflationary Universe, cosmological phase transitions, and the nature of cosmic background radiation fields.

Chapter 4: Stellar Explosions, Neutron Stars, and Black Holes
ALEXEI V. FILIPPENKO

A Professor of Astronomy at the University of California, Berkeley, Alexei Filippenko received his undergraduate training in physics at the University of California, Santa Barbara (1979) and his Ph.D. in astronomy from the California Institute of Technology (1984). Editor of two major monographic works in astronomy, he is also author of more than 350 scientific contributions dealing with supernovae and supernova remnants, quasars and active galactic nuclei, starburst galaxies, and the breakthrough technology of robotic telescopes. Dr. Filippenko has been awarded both a National Science Foundation Young Investigator Award and the Newton Lacy Pierce Prize of the American Astronomical Society. A superb teacher, he is recipient of his university's Distinguished Teaching Award and of the Donald S. Noyce Prize for Excellence in Undergraduate Teaching.

Chapter 5: The Origin and Evolution of the Chemical Elements
VIRGINIA L. TRIMBLE

A *magna cum laude* graduate of UCLA, Virginia Trimble received her advanced degrees in astronomy and physics from Cambridge University, England (M.A.), and the California Institute of Technology (M.S., Ph.D.). A Professor of Physics at the University of California, Irvine, and a Visiting Professor at the University of Maryland, she has also served as a Distinguished Visiting Scholar at the University of Texas, Utah State University, and at Swarthmore College and Mt. Holyoke College. A prolific author, she has published more than 270 contributions dealing with diverse aspects of astronomical science. Dr. Trimble is a Founding Member of the European Astronomical Society, a Fellow of the American Association for the Advancement of Science and the American Physical Society, and recipient of the Outstanding Young Scientist Award of the Maryland Academy of Sciences (1976) and the J. Murray Luck Prize of the National Academy of Sciences (1986).

Biographical Sketches

Chapter 6: The Origin and Evolution of Life in the Universe
CHRISTOPHER P. MCKAY

After receiving a Ph.D. degree in astrogeophysics from the University of Colorado in 1982, Christopher McKay joined the staff at the NASA Ames Research Center in northern California where he serves as a Research Scientist. His current studies focus on understanding the relationship between the origin of life and the chemical and physical evolution of the solar system. He is also actively involved in the planning of future missions to Mars (including those designed to establish human settlements on that planet) to search for evidence of possible past Martian life. For more than a decade, Dr. McKay has been involved in polar research, and has traveled to the Antarctic dry valleys and, recently, to the Siberian Arctic, to conduct research on the forms of life that inhabit these harsh, Mars-like settings. He has strong interest in involving students in the planning of space exploration, especially exploration of Mars.

Chapter 7: Future of the Universe
ANDREI LINDE

Born in Moscow and educated in the former Soviet Union, Andrei Linde graduated from Moscow State University in 1972. His doctoral dissertation, completed in 1974, dealt with the theory of cosmological phase transitions — a series of events that, in accordance with unified theories of weak, strong, and electromagnetic interactions, are thought to have occurred during the very early stages of cosmic evolution. For this work, widely regarded as a benchmark in the development of cosmological theory, he was awarded the Lomonosov Prize of the USSR Academy of Sciences. Dr. Linde is also one of the principal authors of inflationary cosmology, a theory that seems to be more general than (and free from certain of the internal problems of) the Big Bang theory of the origin of the Universe. After working for two years in Geneva, Switzerland, at the internationally acclaimed particle research facility, Conseil Européen pour la Recherche Nucleaire (CERN) (now called the Organisation Européenne pour la Recherche Nucleaire), in 1989 he joined the faculty at Stanford University, where he is currently Professor of Physics.

CHAPTER 1

THE ORIGIN OF THE UNIVERSE

Edward L. Wright

INTRODUCTION

One of the most significant developments of twentieth-century natural philosophy has been the acquisition, through astronomical observations and theoretical physics, of a substantial understanding of the earliest moments in the history of the Universe. The best current model for the origin of the Universe is known as the **inflationary scenario** in the **Hot Big Bang** model. This chapter describes the observational underpinnings of the Big Bang theory and some theoretical models based on it. The development uses elementary mathematics because this is the simplest way to describe much of physical science.

EXPANSION OF THE UNIVERSE

The discovery by Hubble (1929) that distant galaxies are moving away from us with a velocity, v, that is proportional to their distance, D, was the first evidence for an evolving Universe. Observations show that this recessional velocity is

$$v = H_o D \pm v_{pec} \tag{1}$$

where the coefficient H_o is known as the *Hubble constant* and v_{pec} is the peculiar velocity of the galaxy, which is peculiar in the sense that it is different for each galaxy. Typical values of v_{pec} are 500 km/sec, and the recessional velocities v range to a few times 100,000 km/sec for the most distant visible objects.

The Hubble law does not define a center for the Universe, although all but a few of the nearest galaxies seem to be receding from our Milky Way galaxy. Observers on a different galaxy, A, would measure velocities and distances relative to themselves, so we replace v by $v - v_A$ and D by $D - D_A$. The Hubble law for A is

$$v - v_A = H_o(D - D_A) \tag{2}$$

which is exactly the same as the Hubble law seen from the Milky Way (except for slightly different peculiar velocities), provided that $v_A = H_o D_A$. Thus, every galaxy whose velocity satisfies the Hubble law will also observe the Hubble law. An observer on any galaxy sees all other galaxies receding; thus, the Big Bang model is "omnicentric."

The idea that the Universe looks the same from any position is codified in:

> The Cosmological Principle: The Universe is **homogeneous** and **isotropic**.

Note that the Hubble law does not define a new physical interaction that leads to an expansion of the Universe. Instead, it is only an empirical statement about the observed motions of galaxies. Each individual object moves on a path dictated by its initial trajectory and the forces that act on it. Since a uniform distribution of matter produces no net force by symmetry, the only forces felt by an object are caused by the structures around it, and these forces can be computed to good accuracy using the equations of Newtonian mechanics and gravity. On small scales, local forces dominate the motions of objects. For example, the orbit of an electron in an atom is determined by electrostatic forces, and the distance between the electron and the nucleus does not increase with time. The orbits of the planets in our solar system are determined by the gravitational force of the Sun, and the distance between a planet and the Sun does not follow the Hubble law. Note that the individual galaxies in Color Plate 1.1 do not expand. But the motion of galaxies more distant than 30 million light-years is well described by the Hubble law, and this observed fact tells us that the density distribution in the early Universe was almost uniform and that the initial peculiar velocities were small.

The recessional velocity of an object is easily measured with use of the **Doppler shift**, which causes the length of electromagnetic waves received from a receding object to be larger than the wavelength at which they were emitted by the object. Because the long wavelength end of the visible spectrum is red light, the Doppler shift of a receding object is called its **redshift**. The observed wavelength λ_o is larger than the emitted wavelength λ_e, and the ratio

$$\lambda_o/\lambda_e = 1 + z \tag{3}$$

defines the redshift z used by astronomers. For velocities that are small compared with the speed of light, the approximation $v = cz$ can be used. The most distant known **quasar** has $z = 4.9$. For this object, the ultraviolet Lyman α line of hydrogen with $\lambda_e = 122$ nm is seen at $\lambda_o = 720$ nm in the near infrared. For such large z, relativistic corrections to the Doppler shift formula are needed. Color Plate 1.2 shows three galaxies: a bright nearby galaxy, with **magnitude** = 9, a galaxy 100 times fainter (and hence 10 times farther away) with magnitude = 14, and a faint galaxy another factor of 100 times fainter. As predicted by the Hubble law, the recessional velocity increases by a factor of 10 for each factor of 100 decrease in brightness.

The Hubble law in Equation 1 applies to the relative velocity between any pair of galaxies. For example, the velocity of galaxy A with respect to galaxy B is $v_{AB}(t_o) = H_o D_{AB}(t_o)$, where $D_{AB}(t_o)$ is the separation now (the time "now" is denoted t_o) between galaxies A and B. If we consider the separation between A and B after a small time interval Δt, it is

$$D_{AB}(t_o + \Delta t) = D_{AB}(t_o) + v_{AB}\Delta t = D_{AB}(t_o)(1 + H_o\Delta t) \tag{4}$$

The time interval Δt must be a small fraction of the age of the Universe, and yet the distance light travels in Δt must be larger than structures like clusters of galaxies in which local forces produce large peculiar velocities. Observations of the Universe show that it is smooth enough on medium-to-large scales for Equation 4 to be valid. The factor $(1 + H_o\Delta t)$ is independent of which pair of galaxies A and B is chosen, so it represents a universal scale factor that describes the expansion of every distance between any pair of objects in the Universe. This means that the patterns of galaxies in the Universe retain the same shape while the Universe expands, as seen in Color Plate 1.1. We call the universal scale factor $a(t)$, so

$$a(t_o + \Delta t) \approx (1 + H_o\Delta t) \tag{5}$$

for times close to the present. Note that $a(t_o) = 1$ by definition.

If there is no acceleration because of gravity, objects will move with constant velocity and Equation 5 is true even if Δt is not small. In this case, when $\Delta t = -1/H_o$, $a(t_o + \Delta t) = 0$, where a negative Δt denotes an epoch earlier than the present. Thus, all distances in the Universe go to zero at a time $1/H_o$ ago. We normally simplify discussions by defining the moment with $a(t) = 0$ (the "Big Bang") to be $t = 0$. This definition makes the age of the Universe equal to the current time, t_o. For the no-acceleration case, $a(t) = t/t_o$ and the product of the Hubble constant and the age of the Universe is $H_o t_o = 1$. This implies that observers who lived earlier in the history of the Universe, with a smaller t_o, would find a larger Hubble constant H_o. Thus, the Hubble constant is not a physical constant like the electron charge e, because, although the Hubble constant is the same everywhere in the Universe, it changes with time. We call this changing value $H(t)$ and define $H_o = H(t_o)$.

The exact formula for the redshift of an object is $1 + z = a(t_o)/a(t_e)$, where t_e is the time the light was emitted. This states that wavelengths of light expand by exactly the same scale factor that applies to the separations between pairs of galaxies.

The acceleration caused by gravity vanishes only if the Universe is empty, with no mass. When masses are present, gravity provides an attractive force that causes the expansion to slow down. This means that velocities were greater in the past; thus, for a given expansion rate now (H_o), the time since $a = 0$ is smaller. In the most likely case, the density of the Universe is very close to the **critical density** that divides underdense universes that expand forever from overdense universes that will eventually stop expanding and recollapse.

When an small object of mass m is moving under the influence of gravity near a large mass M, the equation that relates its velocity and distance is

$$E = \tfrac{1}{2}mv^2 - \frac{GMm}{r} \tag{6}$$

where E is the total energy, which is conserved, $\tfrac{1}{2}mv^2$ is the kinetic energy, and $-GMm/r$ is the gravitational potential energy. We can use this simple equation in cosmology, with m being a galaxy and M being the mass of the Universe within radius r, which is the density ρ times the volume of a sphere $4\pi r^3/3$. The sphere is centered at $r = 0$, and the galaxy m is located on its surface. (*Proving* that we can use this equation requires general relativity.) Because all matter at larger distances than r has larger velocities than $H_o r$, the matter outside the sphere stays outside. Newton showed that the gravitational force on m from matter outside the sphere is zero, and this is still true under general relativity. Because all matter at smaller distances than r has smaller velocities than $H_o r$, the matter inside the sphere stays inside. Thus, the mass of the sphere is constant. For a body to just barely escape to $r \to \infty$ requires a total energy $E = 0$. This gives the formula for the **escape velocity**, $v_{esc} = \sqrt{2GM/r}$. When the Universe has the critical density, the Hubble velocity $H_o r$ is equal to the escape velocity, which gives an equation for the mass M leading to the critical density:

$$\rho_{crit} = \frac{3H_o^2}{8\pi G} \tag{7}$$

When the Universe has the critical density, we can find the form for $a(t)$ by considering the motion of a galaxy with distance $r = R$ now. At an earlier time, t, its distance was $a(t)R$, so we can estimate its velocity using $v \approx a(t)R/t$, and this is always equal to the escape velocity; thus, $a(t)R/t \approx \sqrt{2GM/(a(t)R)}$. This gives $a^{1.5}/t = $ constant, and since $a(t_o) = 1$, we obtain $a(t) = (t/t_o)^{2/3}$ and $H_o t_o = 2/3$. The age of the Universe is probably between 10 and 20 Gyr, or $t_o = (3.2 \text{ to } 6.3) \times 10^{17}$ sec. Since theory indicates that $H_o t_o$ should be between 2/3 and 1, we expect H_o to be in the range $(1.1 \text{ to } 3.2) \times 10^{-18}$ sec^{-1}. To have a more convenient scale for H_o, the mixed units of km/sec/Mpc are commonly used. A **parsec** is 3.26 light-years, or 3.08×10^{13} km; a megaparsec (Mpc) is 3.08×10^{19} km. Thus, H_o should be in the range 34 to 102 km/sec/Mpc. Observations with the Hubble Space Telescope (HST) by Freedman et al. (1994) give $H_o = 80 \pm 17$ km/sec/Mpc, but other HST observations by Saha et al. (1994) give $H_o = 52 \pm 9$ km/sec/Mpc. The Freedman et al. determination is based on measurements of **Cepheid** variable stars in the large spiral galaxy M100 and agrees with many recent determinations of H_o based on Cepheids; Saha et al. measured Cepheids in a dwarf galaxy. The higher measurement for the Hubble constant, $H_o = 80$ km/sec/Mpc, and the most likely age of the Universe, $t_o = 14$ Gyr, together give $H_o t_o = 1.15$, which is not consistent with the relation $H_o t_o = 2/3$ for a critical density Universe.

One solution to this problem would be to hypothesize that the expansion of the Universe is accelerating instead of decelerating. This hypothesis requires something that acts like antigravity on large scales, and the **cosmological constant** introduced by Einstein to cancel gravity in his early model of a static Universe model could provide the required effect. But because the Universe is not static, the cosmological constant has been regarded as an unnecessary complication by most cosmologists.

The greatest difficulty in cosmology today is in determining the true distances to objects, as opposed to simply using their recessional velocities in the Hubble law. But to measure the Hubble constant, true distances as well as recessional velocities must be measured. Hubble tried this in 1929, but the distances he used were 5 to 10 times too small, and his value for H_o was 5 to 10 times too large. This gave an age for the Universe of $t_o = 1.3$ to 2.0 Gyr, which was less than the well-known age of the Earth. Hubble's error in measuring distances led to the development of the **Steady State** model of the Universe, in which $a(t) = \exp(H_o(t - t_o))$. The Steady State model has an accelerating expansion and a large effective cosmological constant. Because $\exp(H_o(t - t_o)) \to 0$ only for $t \to -\infty$, the Steady State model gives an infinite age for the Universe. However, the Steady State model made definite predictions about the expected number of faint radio sources, and observations made during the 1950s showed that the predictions were wrong. The moral of the Steady State story is that we should not worry too much about the apparently high value of $H_o t_o$ until both the Hubble constant and the age of the Universe are much better known.

The critical density is very low — only 6 hydrogen atoms per cubic meter for $H_o = 75$ km/sec/Mpc. A very good laboratory vacuum (10^{-10} torr) has 4×10^{12} atoms per cubic meter. While the critical density is low, the apparent density of the mass contained in visible stars in galaxies, when smoothed out over all space, is at least 10 times smaller! Thus, the Universe appears to be underdense, which means that E in Equation 6 is positive and the Universe will expand forever. However, this situation is unstable. Consider what will happen as the Universe gets 10 times older. If the density is really only 10% of the critical density now, the Universe will expand at essentially constant velocity, and thus will become 10 times larger. As a result, the density will become 1000 times smaller, since the same amount of matter is spread over 10^3 times more volume. The critical density will also change because the Hubble constant H is a function of time. When the Universe is 10 times older, the value for H will be approximately 10 times smaller. This gives a critical density that is 100 times smaller than the present density. Thus, the ratio of density to critical density becomes 1%. But we can start our calculations of the Universe when $t \approx 10^{-43}$ sec, and $t_o \approx 10^{18}$ sec. If the density were 99% of the critical density at $t = 10^{-43}$ sec, it would be 90% of the critical density at $t = 10^{-42}$ sec, 50% of the critical density at $t = 10^{-41}$ sec, 10% of the critical density at $t = 10^{-40}$ sec, and so on. For the actual density to be between 10% and 200% of the critical density now, the ratio of density to critical density had to be

$$0.999999999999999999999999999999999999999999999999999 < \frac{\rho}{\rho_{crit}} < 1.000000000000000000000000000000000000000000000000001 \quad (8)$$

at $t = 10^{-43}$ sec. This ratio ρ/ρ_{crit} is known as Ω, and we see that Ω has to be almost exactly 1 early in the evolution of the Universe. Figure 1.3 shows three scale factors computed for three slightly different densities 10^{-9} sec after the Big Bang. The middle curve has the critical density of 447 sextillion gm/cm^3, but the upper curve is a universe that had only 1 gm/cm^3 of 447 sextillion gm/cm^3 less density and now has a density lower than the observed density of the Universe; the lower curve is a universe that had 1 gm/cm^3 more and is now at the "Big Crunch." To get a universe like the one we see requires either very special initial conditions or a mechanism to force the density to equal the critical density. Any physical mechanism that sets the density close enough to the critical density to match the present state of the Universe will probably set the actual density of the Universe to precisely equal the critical density. But most of the density in the Universe cannot be stars, planets, plasma, molecules, or atoms. Instead, most of the Universe must be made of **dark matter** that does not emit light, absorb light, scatter light, or interact with light in any of the ways that normal matter does, except by gravity.

COSMIC BACKGROUND

Penzias and Wilson (1965) reported the discovery of a microwave background with a brightness at a wavelength of 7 cm equivalent to that radiated by an opaque, nonreflecting object with a temperature of 3.7 ± 1 **kelvins** (K). Further observations at many

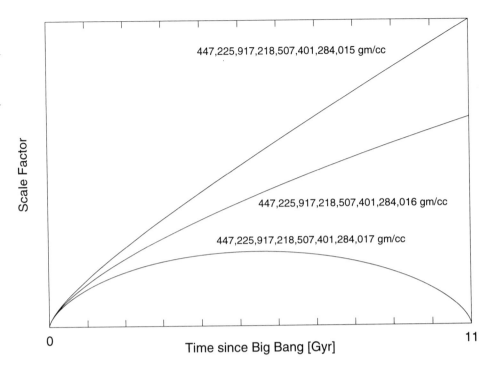

FIGURE 1.3

Scale factor $a(t)$ for three different values of the density of the Universe at $t = 10^{-9}$ seconds after the Big Bang. Note how a very tiny change in the density produces huge differences now.

wavelengths from 0.05 to 73 cm show the brightness of the sky is equivalent to the brightness of an opaque, nonreflecting object (a **blackbody**) with a temperature of $T_o = 2.726 \pm 0.005$ K. The spectrum of the sky as a function of wavelength differs from an exact blackbody spectrum by less than ± 100 parts per million. This shows that the Universe was once very nearly opaque and very nearly isothermal. By contrast, the Universe now has galaxies scattered about and separated by vast stretches of transparent space. Because the conditions necessary to produce the microwave background radiation are so different from the current conditions, we know that the Universe has evolved a great deal over its history. Since the Steady State model predicted that the Universe did not evolve, its predictions are not consistent with the observed microwave background.

Observations of the microwave background toward different parts of the sky show a small variation in the temperature, with one side of the sky being 3.3 mK hotter and the opposite side of the sky being 3.3 mK colder than the average. The pattern of a hot pole and a cold pole is called the **dipole**. It is a measure of the peculiar velocity of our solar system at 370 ± 3 km/sec relative to the Hubble law. The velocity is the sum of the motions caused by the revolution of the Sun around the Milky Way, the orbit of the Milky Way around the center of mass of the local group of galaxies, and the motion of the local group caused by the gravitational forces from the **Virgo Supercluster**, the **Great Attractor**, and other clumps of matter. The local group contains about 30 galaxies, of which the Milky Way and the Andromeda nebula are the biggest; the Virgo Supercluster contains thousands of galaxies.

After the dipole pattern is accounted for, the remaining temperature fluctuations are very small, only 11 parts per million. These tiny temperature differences were detected by the COsmic Background Explorer (COBE) satellite. This implies that the initial density fluctuations in the Universe were also very small.

The current energy density of the microwave background is quite small, as might be expected for the thermal radiation from something that is colder than liquid helium. The number density of microwave photons is 410 cm^{-3}, and the average energy per photon is 0.00063 eV. Thus, the energy density is only 0.26 eV/cm^3. This is 10,000 times less than the critical density. But when the Universe was very young (e.g., $t = 10^{-6}t_o$, about 10,000 years after the Big Bang) and the scale factor was very small, $a(t) = 10^{-4}$, the number density of photons was much greater, 4.1×10^{14} cm^{-3}, and the energy per photon was also much greater, 6.3 eV. The photon density and average energy per photon correspond to a hotter blackbody with a temperature $T = T_o/a(t) = 27,260$ K. As the Universe expands, it also cools. Thus, the energy density of the background when the Universe was small, dense, and hot was very large, 2.6×10^{15} eV/cm^3 when $a(t) = 10^{-4}$, and dominated the density of the Universe for all times less than 10,000 years after the Big Bang.

The very small temperature fluctuations show that the density had fluctuations of about 33 parts per million 10,000 years after the Big Bang. Once the energy density of background radiation becomes less than the density of matter, the fluctuations grow as the denser regions gravitationally attract more material. The process of gravitational collapse makes the fluctuations grow in proportion to the scale factor $a(t)$. Thus, in the case of a Universe with the critical density described above, the Universe was no longer radiation-dominated when $a = 10^{-4}$, so the fluctuations grew from 33 parts per million to 33%. This is just enough to explain the observed clustering of galaxies that we see in the Universe now. But if the Universe had no dark matter, then $a = 10^{-3}$ when the Universe stopped being radiation-dominated. Furthermore, ordinary matter interacts with light and could not move through the background radiation until the Universe was cold enough for neutral hydrogen atoms to be stable. This happened approximately 300,000 years after the Big Bang, when the temperature of the microwave background fell to 3000 K, which, coincidentally, is when $a = 10^{-3}$. Thus, if only ordinary matter had been present, the fluctuations implied by the COBE observations would have grown only to 3.3% at the current epoch, much less than the presently observed clustering of galaxies.

LIGHT ELEMENT ABUNDANCES

Although the energy density of the microwave background dominated the Universe for the first 10,000 years after the Big Bang, it is even more significant during the first 3 minutes. One second after the Big Bang, the average energy of a photon was 3 MeV, which is a **gamma ray**. Gamma rays destroy any atomic nucleus; thus, 1 second after the Big Bang there were only protons (p or hydrogen nuclei), free neutrons (n), electrons (e^-) and positrons (e^+), neutrinos (ν_e, ν_μ, and ν_τ), and antineutrinos ($\overline{\nu}_e$, $\overline{\nu}_\mu$, and $\overline{\nu}_\tau$). The three types of neutrinos correspond to the three generations of elementary particles, but all the particles in the second and third generations, except for the neutrinos, are so heavy and unstable that they decay during the first second after the Big Bang. Weak nuclear interactions such as

$$p + e^- \leftrightarrow n + \overline{\nu}_e \qquad (9)$$

determined the ratio of neutrons to protons. Because the neutron is heavier than the proton, the neutron-to-proton ratio declines as the temperature falls. But eventually, at about 1 second after the Big Bang, the density of electrons and neutrinos falls to such a low level that reaction 9 is no longer effective. After this time, the neutron-to-proton ratio gradually falls because of the radioactive decay of the neutron,

$$n \rightarrow p + e^- + \overline{\nu}_e \tag{10}$$

which has a half-life of 615 seconds.

As neutrons decay, the Universe expands and grows colder. Eventually the temperature falls to the point where the simplest nucleus, the heavy hydrogen or deuterium nucleus (d), is stable. This occurs when the temperature is about 10^9 K, which occurs about 100 seconds after the Big Bang. At this point, the reaction

$$p + n \rightarrow d \tag{11}$$

very quickly converts all neutrons into deuterium nuclei. Once deuterium is formed, it is quickly converted into helium through a network of interactions, with the net effect

$$d + d \rightarrow He \tag{12}$$

Because almost all neutrons that survive until T is less than 10^9 K end up bound in helium nuclei, the helium abundance in the Universe provides a measurement of the time it takes for the Universe to cool to 10^9 K. If the Universe cools rapidly, there is a large helium abundance, but slow cooling gives low helium abundance because more of the neutrons decay. The standard Big Bang model, with three types of neutrinos, predicts a helium abundance that is correct to within the 1% margin of uncertainty.

Reaction 12 requires collisions between two nuclei, and if the density of atomic nuclei is low, then a fraction of the deuterium will not react. Thus, the residual fraction of deuterium in the Universe is a sensitive measure of the density of atomic nuclei. Based on the abundance of deuterium and other light isotopes like ^3He, the best estimate for the current density of nuclei of all sorts is equivalent to 1/6 hydrogen atoms per cubic meter (Copi, Schramm, and Turner, 1995). This is about 40 times less than the critical density. Because the density of the Universe must be close to the critical density to produce the observed clustering of galaxies, we find from the light element abundances that more than 90% of the mass of the Universe must be the mysterious dark matter.

In the 1940s, Gamow and colleagues proposed that all the chemical elements were produced in the Big Bang. This proposal led to a prediction of a 5-K microwave background (Alpher and Herman, 1948), but this prediction was not followed up. The eventual discovery of the microwave background in 1964 was accidental. Why was this prediction ignored? The absence of stable nuclei with atomic weights of 5 and 8 means that the Big Bang produces only hydrogen and helium isotopes and a very small amount of lithium. When a model is supposed to produce all the elements from $Z = 1$ to 92, but actually only works for $Z = 1$, 2, and 3, its other predictions tend to be ignored. But in this case the predictions were right.

HORIZONS

We can see only a finite piece of the Universe. The naive estimate for how far we can see is ct_o, the speed of light times the age of the Universe. This is, in fact, the distance traveled by photons coming from the most distant visible parts of the Universe, as measured by the photons. But when one defines distances in an expanding universe, the convention is to measure all intervals at the current time, t_o. Because the Universe has expanded since $t = 0$, the earlier parts of the photon's journey get extra credit. We can compute the distance we can see in a critical density universe by dividing the age of the Universe into more and more intervals. With one interval, we get ct_o. With two intervals, we get $0.5ct_o/0.5^{2/3} + 0.5ct_o = 1.29ct_o$, because the first half of the journey has expanded by the factor $1/a(t_o/2) = 1/0.5^{2/3}$. With four intervals, we obtain $ct_o(0.25/0.25^{2/3} + 0.25/0.5^{2/3} + 0.25/0.75^{2/3} + 0.25) = 1.58ct_o$. With a very large number of intervals, we get $3ct_o$, which is the distance to the **horizon**. For $t_o \approx 13$ Gyr, this is 40 billion light-years.

Consider now an observer 300,000 years after the Big Bang. The distance to the horizon is $3ct$, or about 1 million light-years. This observer (really just a cloud of gas) will try to get in thermal equilibrium with the region it can see, which extends to a 1-million-light-year radius. If thermal equilibrium can be achieved, a patch of constant temperature 1 million light-years in radius can be created. This patch will grow to 1 billion light-years in radius as the Universe expands from 300,000 years after the Big Bang until now. But our horizon now is 40 billion light-years in radius. Thus, the constant temperature patch subtends an angle of only 1/40 radian, which is only three times the diameter of the full moon. But we see an almost constant temperature over the entire sky. For a universe to be as isotropic as the one we live in requires either very special initial conditions or a mechanism to force the temperature to be constant over the entire observable Universe.

INFLATION

The Big Bang model described above is in good agreement with the observed Universe, but it required very special initial conditions to explain two different facts:

1. The fact that $\rho/\rho_{crit} = \Omega$ is close to 1 today means that Ω was nearly exactly 1 initially.
2. The microwave background temperature is nearly identical in patches that could not have communicated with each other before the Universe became transparent 300,000 years after the Big Bang.

In 1981, Guth proposed the **inflationary scenario**, which attempts to make these initial conditions less special (see Guth and Steinhardt, 1984, in Further Reading). The *inflationary scenario* supposes that at some time during the early history of the Universe, a large cosmological constant existed, which led to a rapidly accelerating expansion of the Universe. In Russia, Starobinsky (1979) began to study universes in which a rapidly accelerating expansion preceded the normal decelerating expansion of the Big Bang. During this inflationary epoch, the Universe was like the Steady State model, but only temporarily. The prologue of the inflationary scenario is a normal Big Bang expansion until a time t_s. In Linde's model of perpetual inflation (see Chapter 7), this prologue is absent because the Universe begins during inflation. At t_s after the Big Bang, Act I, the inflationary phase, begins. During this time, the Hubble constant is $H \approx 0.5/t_s$. During the inflationary epoch, the Universe expands by a factor of 10^{43} or more. At about $200t_s$, the inflationary epoch ends, and Act II begins. Act II is a standard Big Bang model, but with initial conditions set during the inflationary epoch. For example, a very small patch can become isothermal, and inflation makes the small isothermal patch into a huge isothermal patch, which expands to become much bigger than the observable Universe.

But why does inflation make $\Omega \to 1$ almost exactly? The answer lies in the Steady State nature of the inflationary epoch. When the Universe expands, one expects the density to go down, but in a Steady State model, the density must remain constant. Thus, there must be a continuous creation of matter during a Steady State epoch. This means that the mass M in Equation 6 gets bigger, like r^3, instead of staying constant. Then the potential energy term grows increasingly negative as the Universe inflates, in proportion to r^2. To conserve energy, the kinetic energy term $mv^2/2$ must get bigger, so v gets bigger, and the expansion accelerates as expected in a Steady State situation. As noted in the section on the expansion of the Universe, this acceleration is equivalent to introducing Einstein's cosmological constant. But note that

$$\frac{GMm/r}{0.5mv^2} = \frac{\rho}{\rho_{crit}} = \Omega \qquad (13)$$

Therefore, if before inflation the Universe had $0.5mv^2 = 2$, and $GMm/r = 1$, so $E = 0.5mv^2 - GMm/r = 1$, and $\Omega = 0.5$, then after inflating by a factor of 10^{43}, we have $GMm/r = 10^{86}$. Then $0.5mv^2$ must be $10^{86} + 1$ to preserve $E = 1$. Thus, after inflation, $\Omega = 1 - 10^{-86}$, which is well within the tight limits given in Equation 8.

Thus, inflation solves two problems in the Big Bang model, but creates another question: why does the Universe have a large cosmological constant during the inflationary epoch? The answer to this question lies in high energy particle physics, under the topic of unified field theories. The Weinberg-Salam model that unifies the **electromagnetic** and **weak nuclear** interactions into a single **electroweak** theory requires a large **vacuum energy density**. A vacuum energy density acts just like a cosmological constant, and in the Weinberg-Salam theory, the Universe makes a **phase transition** from a state with a large cosmological constant when T is greater than 10^{15} K (or when the energy density is equivalently high) to the normal state with zero cosmological constant at lower temperatures. A similar unification of the **strong nuclear** force with the electroweak force gives a **grand unified theory**, or GUT. In GUTs, the transition from high-to-low cosmological constant occurs when T is greater than 10^{28} K. Either of these phase transitions could cause an inflationary epoch.

Inflation produces such a tremendous enlargement that even tiny objects such as the **quantum fluctuations** that occur on subatomic scales get blown up to be the size of the observable Universe. But while the fluctuations are being inflated, new small ones are always being created. Because the time for the Universe to double in size is constant during inflation, the power in the fluctuations in each factor of two size bins is constant. Let us measure time in units of the doubling time, and size in units of the speed of light times the doubling time. Then at $t = 10$, the fluctuations created between $t = 9$ and $t = 10$ are all about size $r = 1$ because they have existed for less than 1 doubling time. At $t = 1$, there should have been the same amount of fluctuations at size $r = 1$. But these fluctuations now have size $r = 512$ and $t = 10$. Hence, at $t = 10$, the amount of fluctuations at $r = 512$ and $r = 1$ should be the same. The same argument, applied at $t = 2, 3, 4 \ldots$, shows that amount of fluctuations at sizes $r = 256, 128, 64 \ldots$ should all be equal to the amount at $r = 512$ and $r = 1$.

These fluctuations become temperature variations, and the equality of the amount of variations on different angular scales is a prediction of the inflationary scenario. In 1992, the COBE team announced the discovery of temperature variations with a pattern that is consistent with equal variations in angular size bins centered at 10°, 20°, 40°, and 80°. Color Plate 1.4 compares a predicted sky map produced using equal power on all scales to the actual sky map measured by COBE. The two maps look quite similar, and a detailed statistical comparison shows that the equal power on all scales prediction of inflation is quite consistent with the observations.

CURRENT RESEARCH

Dark Matter

The small density fluctuations indicated by the small temperature differences seen by COBE can grow into the galaxies and clusters of galaxies that we see in the Universe today, but only if the action of gravity is not impeded by other interactions. The most important epoch for the growth of structures is the period just after 10,000 years after the Big Bang. At this point the density of matter becomes larger than the density of the background radiation, which allows dense regions to collapse under the influence of their own gravity. The temperature differences measured by COBE are a direct indication of the gravitational potential differences, which are equivalent to the heights and depths of mountains and valleys on Earth. In fact, a typical gravitational potential difference corresponds to ±300 million km in a constant gravitational acceleration equal to Earth's surface gravity. But the distance between peaks and valleys in the Universe is astronomical: 300 quadrillion km. Thus, the gradient is very gentle, and only matter that moves freely downslope will be able to gather together in pools in the valleys. All chemical elements are ionized at the temperature of 30,000 K that existed 10,000 years after the Big Bang, and the resulting free electrons interact with the background radiation to produce a very strong interaction that resists the force of gravity. Thus, all ordinary matter acts like molasses and does not flow freely down the small gravitational gradients in the Universe. Therefore, most of the mass of the Universe must be made of exotic material that does not interact with radiation. It can not scatter light, absorb light, or emit light. This is **nonbaryonic** dark matter. The nature of this dark matter is still quite uncertain.

Historically, the first candidate for nonbaryonic dark matter was the neutrino. Neutrinos are known to exist, and their number density, determined by the reactions in Equation 9, is fixed by the observed microwave background. If one of the three kinds of neutrino had a mass about 10,000 times smaller than the mass of the electron, the density of neutrinos in the Universe would be sufficient to give $\Omega = 1$. But neutrinos with this tiny mass would have a speed of about 200,000 km/sec at the critical time 10,000 years after the Big Bang. Because of this rapid motion, neutrinos are called **hot dark matter**. They would thus move about 7000 light-years before slowing down as the Universe expanded. The 7000 light-years would grow to be 70 million light-years now. In any dense region smaller than this, the neutrinos would escape before the dense region could collapse, so neutrino dark matter would produce only very large scale structure.

The next model for nonbaryonic dark matter assumes the existence of a new, heavy, electrically neutral and stable particle. This particle would interact very weakly with ordinary matter and radiation, so it received the name Weakly Interacting Massive Particle (WIMP). Because such a heavy particle would be moving very slowly 10,000 years after the Big Bang, WIMPs are a form of **cold dark matter**. One possible candidate for the dark matter is the **neutralino**, which is the lightest supersymmetric particle. Supersymmetric grand unified theories (Susy GUTs) are a currently favored class of models for the high energy particle interactions observed in particle accelerators. However, the cosmological predictions of the cold dark matter model do not depend on the nature of the WIMP, but they do depend on the quantity ΩH_o. If ΩH_o were equal to 30, the CDM model would match the observations of large scale structure. If ΩH_o is larger than 30 km/sec/Mpc, the cold dark matter model gives too much small scale structure. But because Ω should be equal to 1, and H_o appears to be 2 to 3 times larger than 30 km/sec/Mpc, cosmologists have been forced to consider more complicated models.

One such model mixes hot and cold dark matter (jokingly referred to as "hot and cold running dark matter"). The neutrino mass is assumed to be about 100,000 times smaller than the mass of the electron, since most of the density of the Universe is postulated to be cold dark matter. The mixture of cold dark matter, which gives too much small scale structure, and hot dark matter, which gives no small scale structure, makes a porridge that is "just right."

A second class of models adds a cosmological constant to the cold dark matter (see the section on the expansion of the Universe and the subsection that follows it). This model can also fix the conflict between the large age of the oldest stars and high values of the Hubble constant. However, a cosmological constant is a somewhat more arbitrary addition to the model than a small neutrino mass. After all, we know that neutrinos exist.

Age and the Hubble Constant

The age of the Universe has been estimated from the age of the oldest stars in globular cluster to be 14 to 18 Gyr, while the ages of the oldest white dwarf stars in the solar neighborhood give an estimated age of the Universe of at least 10 to 12 Gyr. Finally, the age of the chemical elements, determined by radioactive decay, is 11 to 17 Gyr. A conservative value for the age of the Universe is thus $t_o = 14 \pm 4$ Gyr. But this age of the Universe would imply a Hubble constant of 46 km/sec/Mpc for $\Omega = 1$, and measurements of Cepheid variable stars in distant galaxies by the HST give $H_o = 80 \pm 17$ km/sec/Mpc. These data give $H_o t_o = 1.15 \pm 0.41$ instead of the 2/3 expected from inflation. However, the uncertainties are large. More data will be obtained, and will either verify the discrepancy, forcing the existence of a cosmological constant, or H_o or t_o will be found to be smaller, which will remove the discrepancy.

Infrared Background

Galaxies first formed about 1 billion years after the Big Bang and are expected to have produced a large amount of infrared radiation during the formation process. Infrared light from billions of distant galaxies all over the sky should produce a uniform infrared glow: the infrared background. The background has not yet been detected because of the interfering radiation produced by material in the solar system and the Milky Way. Color Plate 1.5 shows an infrared picture of the sky produced by the COBE satellite. The bright bar down the equator is infrared emission from our own Milky Way galaxy, and the "lazy S" in the lower left and upper right is infrared emission from the plane of our solar system. The dark spots in the upper left and lower right may contain the infrared background, but a careful subtraction of the infrared emission from our solar system and the Milky Way is needed before the infrared background can be studied.

CONCLUSION

The conclusion of this chapter is merely the introduction to the next act, the origin of galaxies. The conditions necessary to form galaxies were established during an inflationary epoch that occurred earlier than 10^{-12} seconds after the Big Bang. Because inflation makes any preexisting structures in the Universe much too large to be observable, the temperature fluctuations observed by COBE, which were created during inflation, are the oldest structures we can ever observe.

FURTHER READING

Brush, S. G. 1992. How cosmology became a science. *Scientific American 267*: 62.

Freedman, W. L. 1992. The expansion rate and size of the Universe. *Scientific American 267*: 54.

Gulkis, S., Lubin, P., Meyer, S., and Silverberg, R. 1990. The cosmic background explorer. *Scientific American 262*: 132.

Guth, A., and Steinhardt, P. 1984. The inflationary Universe. *Scientific American 250*: 116.

Krauss, L. 1986. Dark matter in the Universe. *Scientific American 255*: 58.

Linde, A. 1994. The self-reproducing inflationary Universe. *Scientific American 271(5)*: 48.

Peebles, P. J. E., Schramm, D. N., Turner, E. L., and Kron, R. G. 1994. The evolution of the Universe. *Scientific American 271(4)*: 52.

REFERENCES

Alpher, R. A., and Herman, R. 1948. Evolution of the Universe. *Nature 162*: 774.

Copi, C., Schramm, D., and Turner, M. 1995. Big Bang nucleosynthesis and the baryon density of the Universe. *Science 267*: 192–199.

Freedman, W., Madore, B. F., Mould, J. R., et al. 1994. Distance to the Virgo Cluster galaxy M100 from Hubble Space Telescope observations of Cepheids. *Nature 371*: 757–762.

Guth, A. 1981. Inflationary Universe: A possible solution to the horizon and flatness problems. *Physical Review (D) 23*: 347–356.

Hubble, E. 1929. A relation between distance and radial velocity among extragalactic nebulae. *Proceedings of the National Academy of Sciences 15*: 168–173.

Penzias, A. A., and Wilson, R. W. 1965. A measurement of excess antenna temperature at 4080 Mc/s. *Astrophysical Journal 142*: 419–421.

Saha, A., Labhardt, L., Schwengeler, H., Macchetto, F. D., Panagia, N., Sandage, A., and Tammann, G. A. 1994. Discovery of Cepheids in IC 4182. *Astrophysical Journal 425*: 14–34.

Starobinsky, A. A. 1979. Relict gravitational radiation spectrum and initial state of the Universe. *Journal of Experimental and Theoretical Physics Letters 30*: 682–685.

CHAPTER 2

THE ORIGIN AND EVOLUTION OF GALAXIES

Alan Dressler

INTRODUCTION

The preceding chapter discussed the Big Bang, the early moments of our Universe when all that we know of existed in a very different form, a swarming sea of ultra-high-energy light and massive particles. In the beginning, this scalding bath of matter and energy was a very smooth one. Today, in contrast, the Universe is cold and dark, a darkness broken by concentrated patches of light — stars, collected into giant star systems we call galaxies. In other words, the Universe has evolved from smoothness to complex structure, an evolution that is inexorably tied to our own existence.

Because of the nature of the Center for the Study of Evolution and the Origin of Life (CSEOL) symposium, I should perhaps be more careful than astronomers usually are when I use the word "evolution" in regard to the Universe. Evolution in this context is not the evolution biologists speak of, which involves some sort of selection process, but rather evolution simply as secular change, as the *Oxford Concise Dictionary* puts it, "appearance (of events, etc.) in due succession." Whether the evolution of the Universe entails any kind of intentionality or design is, of course, beyond the scope of this text, but is fertile ground for the minds of each of us.

It is the progression of the Universe from the great symmetry and simplicity we imagine for the Big Bang, through the building of the particles we see today, with their complicated relationships, through the synthesis of the chemical elements, first in a pervasive hot plasma and then in the cores of stars, and most recently to the invention of biology and life. That is what the phrase "evolution of the Universe" means to me. It is the building of great complexity, best represented by life itself, whose essence is complex variation, from what was utter simplicity. Our own evolution is very much a part of what the Universe has been about through billions of years. We are strongly connected to the evolution of the Universe — we are not a mere sideshow or insignificant accident. This is why it is not vain or arrogant to see as the most important question: What sequence of events led to our own existence?

My chapter in this bit of storytelling is to review what we think we know and what we are still puzzled about concerning the evolution of galaxies, vast systems of billions of stars that are principal building blocks of the Universe at large. Their appearance on the scene and their subsequent evolution is a crucial plot point in the story of creation.

WHAT IS A GALAXY?

When we look into the night sky we see stars, thousands of them, with our unaided eye, and millions if we use even a small telescope. Early in recorded history we find a recognition that many stars are concentrated in a band that wraps around the sky, a "milky way" that, through the Greek words "galaxias kyclos," gave us our English word "galaxy." About 400 years ago, with the newly invented telescope, Galileo Galilei showed that the band is caused by the light of myriads of stars, each too faint to be seen with the eye, but together providing a fuzzy glow. The fact that these stars were collected in a fairly narrow band was correctly guessed to be an expression of the flatness of a vast disk of stars, of which our Sun was just one. More than 200 years ago, intelligent people, such as Immanuel Kant, speculated that there might be many other "island universes" — countless stars collected into vast disks and balls.

In the 1920s Edwin Hubble was an astronomer where I myself work in Pasadena, at the Carnegie Institution, at the Mount Wilson Observatory as it was then called. Hubble put an end to centuries of debate about the reality of "other Milky Ways" when he measured a distance for the fuzzy ellipse in the sky known as the Andromeda nebula (see Figure 2.1). Hubble showed conclusively that this system lay well beyond the boundaries of our Milky Way. He did this by identifying a certain kind of star, a Cepheid variable, that varies in brightness in a regular way. The intrinsic brightness of a Cepheid was known from studies of Cepheids in our Milky Way; knowing how bright a star actually is allows you to calculate how far away it is, by comparing its apparent brightness to its true brightness, and employing the almost familiar rule that the apparent brightness of a luminous source falls as the square of the distance.

From there, Hubble and other astronomers of the day identified which of the fuzzy patches of light in the sky were galaxies and separated them from other smaller associations of starforming regions within the Milky Way, for example, the glowing patch of stars and gas that is the Orion Nebula (Figure 3.1 in Chapter 3). By the late 1920s hundreds of galaxies had been identified. Hubble and others took the light of these galaxies and subjected it to spectroscopic analysis; by breaking the light into its component colors, the combined starlight that made up the vast systems could be compared with the light of our Milky Way and with that of individual stars within the Milky Way. The most startling discovery came from the **Doppler shift** — the same effect we hear as a drop in pitch of the horn from a moving car as it approaches and then passes — which revealed that many of the galaxies are receding from us at speeds of thousands of miles per second! The further a galaxy is from the Milky Way, Hubble found, the greater its speed of recession. The **expansion of the Universe**, a concept Hubble never seemed to feel completely comfortable with, became his most famous legacy and led, eventually, to the notion of a creation event for the Universe — the Big Bang theory — and the realization that the Universe had a dynamic rather than static nature, that it had a past

FIGURE 2.1 Our neighbor, the Andromeda galaxy, a galaxy similar in size and form to our own Milky Way. It lies at a distance of roughly 2 million light-years and is approximately 100,000 light-years in diameter.

different from its present, and a future different still—that the Universe is evolving. These discoveries rank among the greatest in human history because they changed our perception of what the Universe is.

Hubble recognized the different forms in which galaxies are found. Galaxy shapes vary, but the variations are familial, like the difference between a giraffe and a horse, rather than between a giraffe and a jellyfish. The variations come in ranges of size and brightness but represent only scaled versions of what is basically the same animal. This simplicity is repeated over huge distances—a few basic shapes, a wide range in size, a lone galaxy here and there, many doubles and triples, a few groups of several large and many smaller galaxies, perhaps one populous cluster every so often. This is the way the Universe looks—in neighborhoods 100 million light-years in diameter.

Finding the underlying causes of this remarkable regularity is a preoccupation of current astronomical research. As is frequent in science, taxonomy has been a first step—grouping subjects into classes that share some characteristic can uncover clues to the physical process that is responsible for the trait.

Some of the early work in galaxy classification was again done by Hubble. He identified three primary types—**spiral**, **elliptical**, lenticular (lens-shaped)—distinguished by shape and morphology rather than by size or brightness. Examples of these types of galaxies are shown in Figure 2.2. Spiral galaxies, like our Milky Way, are the most common type; they typically appear in the sky as stretched ovals with spiral bands in their outer regions. Astronomers soon recognized that a spiral galaxy is basically a flat disk—those that are seen edge-on are 20 to 30 times as thin as they are across, but those viewed face-on are nearly perfect circles. Those inclined at some intermediate angle to our line of sight are seen as oval-shaped—to be precise, as ellipses.

Not all spiral galaxies are completely flat; many also swell in the middle. This *bulge* is known to be more-or-less spherical because the central regions appear roundish, regardless of the angle at which we view the galaxy. (Only a sphere looks round from every angle.) Even though the view of our own galaxy is obscured by our location in its dust-laden disk, we can see that it has both characteristics—the thinness of the disk is evident as the narrow band we call the Milky Way that rings the sky, and the round bulge of our galaxy shows as a striking widening of this band toward the constellation Sagittarius.

Elliptical galaxies, in contrast, are simply roundish balls of billions of stars. After many years of research, it is still unclear whether they are all egg-shaped (prolate) or onion-shaped (oblate) or whether there are some of both shapes. This has been a difficult issue to settle because we have only one view of the cosmos. We see each galaxy from only one direction. In our lifetimes, instantaneous by cosmic standards, we have no chance to travel to a place from which we might view a galaxy's "better side," and no time to let the galaxy turn beneath our gaze.

Members of the S0 (S-zero) class of galaxies (Hubble called them lenticulars) look like a cross between an elliptical and a thin-disk spiral. The round bulge in the middle dominates the light of the fainter disk that surrounds it, and there is no spiral pattern seen in the disk. There is also a class referred to as "irregular," which is a catchall category for the few percent that are not spiral, elliptical, or S0 galaxies. Some irregulars appear to be galaxies too small to sustain a stable, well-organized shape; others seem to be galactic equivalents of train wrecks, the results of disruptive collisions.

The spiral pattern prevalent in so many galaxies allowed astronomers to move from taxonomy to a rudimentary understanding of the physical processes that give rise to the different types. As more was learned about the life cycles of stars, it became clear that the spiral pattern is delineated by young stars forming in thin disks, and that the

FIGURE 2.2

Different types of galaxies, from the flat disk spirals (*top*) to the round elliptical galaxy (*bottom left*). Spirals are full of gas and dust, the raw materials of ongoing star formation, but elliptical (E) and S0 galaxies are, in comparison, nearly gas-free. (Photos courtesy of the Carnegie Observatories.)

spiral pattern, the thinness of the disk, and the formation of new stars are all intimately related.

Our galaxy is very much typical of all galaxies. It is of the most common type — a spiral — and of typical mass and luminosity. The visible part of our Milky Way galaxy, that is, the area where the stars are, is about 100,000 light-years in diameter, the distance light could travel in 100,000 years — sunlight that bounced off Earth when Neanderthals existed here has not yet crossed the breadth of our galaxy. And our Sun is a fairly typical star, though somewhat more massive than most. There are about 100 billion stars in the Milky Way galaxy, about 20 stars for every human now living on Earth. Our own galaxy is rotating — the Sun and its neighbors are moving at about 150 miles per second around the Milky Way's center. Even at this pace, it will take about 200 million years to complete one orbit — when last the Sun passed this way dinosaurs were just beginning to rule Earth.

We are now certain of the basic picture of what is happening in galaxies. The disks of spiral galaxies are mainly made up of stars moving on near-circular orbits around the galaxy's center. Floating among the stars is a lesser mass of gas from which the stars themselves have formed. The gas is mainly made up of hydrogen atoms. Helium atoms make up most of the rest, but about 1% is all-important heavy atoms, including carbon, nitrogen, oxygen, silicon, magnesium, and iron. Most of the heavy atoms are bound together in tiny grains of hydrocarbons, silicates, and ices — a thin mist of the same material of which Earth and we ourselves are made. The average density of the gas and dust between stars is only one atom per cubic centimeter, a better vacuum than has ever been produced on Earth, but 100,000 times denser than the gas between galaxies, and dense enough to be a starting point for star formation.

Along the spiral arms in a galaxy are thicker clouds of gas where new stars are condensing and lighting their nuclear fires. In at least some cases, the spiral pattern itself seems to be the agent for star formation. A spiral arm is like a pressure wave that sweeps continually around the galaxy, the way water waves slosh back and forth in a bathtub. As it passes, the wave squeezes the cold, dormant gas, compressing it to where the pull of gravity takes over and drives an avalanche-like collapse. As each gas cloud contracts its temperature rises; this causes it to glow more brightly, and the loss of this radiant energy causes the cloud to contract further (see Chapter 3). Only fragmentation into stellar size globules can stall this inexorable collapse, for in this new form central temperatures rise to millions of degrees, igniting the enormous power source of nuclear fusion. The energy released by a star's central nuclear fire releases tremendous heat — in a gas this translates into pressure, enough pressure to counterbalance the staggering weight imposed by gravity.

HOW DID GALAXIES FORM? — THE SHORT VERSION

There is little doubt about the basic reason why galaxies exist in the Universe: gravity is responsible. Gravity is a characteristic of matter — a universal attraction between any and all mass/energy, in direct proportion to the amount and in inverse proportion to the distance squared. We cannot say why there is gravity and do not yet understand fully how the force of gravity is exerted, but its reality is not in question, nor is its role in building galaxies.

We can see back to a time when the Universe had no galaxies, when the matter was very smoothly distributed. What has happened is that gravity has collected huge amounts of matter into vast "continents" separated by a comparatively empty sea; within these continents, gravity has further drawn matter into individual concentrations called stars, where the amazing process of nuclear fusion is underway.

Galaxies formed because of gravity. But why are there variations in form? Why are some galaxies full of young stars outlining spiral arms and others — the elliptical galaxies — round balls of only old stars wending their way to eventual death?

What about the distribution of galaxies themselves? We have learned, especially during the past decade, that galaxies are not strewn randomly through space, but are distributed in lacy patterns and dense clusters, separated by nearly empty voids — more like sculpture than spatter. Surely gravity is responsible for this too, but what is it about gravity working on the initial conditions of the Universe (that is, the physics of the Big Bang) that carves these almost deliberate-looking patterns? Spiral galaxies are more commonly found in low-density regions; elliptical galaxies are found more commonly in denser groups, and most commonly in very dense clusters. Why? What mechanism regulates which kinds of galaxies form, and why do their relative numbers vary in different space environments?

For major galaxies like the Milky Way, a typical range in size is a factor of 10, which corresponds to a range in mass of a factor of 100 to 1000. In most respects, one galaxy can look exactly the same as another, even if their diameters differ, say, by a factor of 5. (Compare this with the variation in the size of adult humans, which is typically a variation of about 20%. If one makes the range include infants through the tallest basketball players, an idea of the range in size that spiral galaxies cover is better visualized.) The variation in galaxy size is believed to be a simple consequence of the size of a ripple in the previously smooth distribution of matter that became the galaxy: large ripple, large galaxy; small ripple, small galaxy. Gravity brought together what was at hand.

What about the existence of spiral and elliptical galaxies and the irregular types that occur more rarely? There are some basic ideas on this, though no one has succeeded in making a convincing argument of why different kinds of galaxies form. One suggestion comes from consideration of *angular momentum*, that is, the degree to which the matter that forms a galaxy is spinning. The notion is that spirals formed from material that was spinning quite a lot; as a result, the stars are generally all orbiting in the same direction. This is a system of high angular momentum, and it explains the flattening of the galaxy. In ellipticals, on the other hand, stars are circulating in all directions — they are on regular orbits, but they line up every which way, and many of them have long, skinny orbits rather than circular ones. These are systems of low net angular momentum.

To understand how angular momentum could determine which kind of galaxy is produced, the question of where the spin comes from in the first place must be answered. The amount of spin tends to dissipate as the Universe expands — by the time galaxies were forming we would not expect much spin to be left in the cloud of material that would form a galaxy. But years ago, Princeton physicist Jim Peebles showed that if two galaxies were forming somewhat near each other, they might exert a gravitational force that would set each of them spinning in opposite directions, and this initial kick would result in a permanent spin if the expansion of the Universe drew the objects apart. In this picture, galaxies that formed near a neighbor that has since moved away could turn into spirals; those in which the two or more neighbors fell back together again and merged (this could have been at a time before we might have recognized them as galaxies) would cancel their spins and return to slowly rotating, that is, elliptical galaxies. This mechanism is likely a part of the puzzle of why some galaxies become spirals and others become ellipticals.

Another factor has to do with the rate of star formation. A galaxy with a prominent disk of gas and dust is one in which the formation of stars did not exhaust all the gas. A galaxy that forms from an enormous gas cloud that is contracting under the pull of gravity, with stars forming in knots within the cloud, may produce a spiral galaxy with its prominent disk, corresponding to the case in which star formation proceeded more slowly than the contraction of the galaxy, so that there was gas left over to make the disk. In contrast, an elliptical galaxy might be a system in which star formation proceeded so rapidly that there was no leftover gas to form a disk by the time the contraction was complete. It has been suggested that the initial density of the system (how much material was packed into a given volume) regulated which process happened faster, the contraction of the system or the complete exhaustion of the gas in star formation. In this model, the ellipticals are thought to be the remnants of the densest gas clouds.

These ideas are supported by discoveries, beginning in the 1940s with Walter Baade's work at the Mount Wilson Observatory, that there are two distinct populations of stars in spiral galaxies like our own. The stars of the spheroid or bulge of a galaxy — the round part — are found to be, on average, much older than the stars in the disk — the flat part of the galaxy. (The techniques by which a **stellar population** can be assigned an average age are considered to be very reliable.) Furthermore, the generally older populations have a greater representation of stars with very low abundances of heavy chemical elements including, for example, carbon, nitrogen, oxygen, silicon, magnesium, and iron. As discussed in Chapter 5, these elements were formed in generations of stars — they were not produced in the Big Bang. The presence of stars with very low proportions of these elements is in keeping with the idea that these stars are some of the oldest in a galaxy, stars born before the wholesale production of heavy chemical elements. This supports the notion that the spheroidal part of a galaxy (or an elliptical galaxy, which is entirely spheroid) is the first to take form in the process of galaxy evolution, and that the stars of the disk, with their higher proportions of heavy elements and generally younger ages, came later.

If these are the kinds of processes that might be at work, how do they explain the prevalence of elliptical galaxies in dense clusters and of spirals in the low-density suburbs of the Universe? This raises the question of nature versus nurture for galaxy formation. Perhaps the initial conditions, like the density of gas clouds in what would become a cluster of galaxies, or the spins the objects generated in each other, account for the difference in populations. In this way, the different types would be set from the very beginning, in a sense, therefore, predetermined by nature. Other astronomers believe that all galaxies were born alike, probably as spiral galaxies, and that elliptical galaxies were made when spirals ran into each other or were otherwise mangled in the dense environment of a cluster.

In any event, it is clear that clusters of galaxies are a good laboratory in which to study galaxy formation because many things could happen in a dense environment that could influence a galaxy. Given the large numbers of relatively rare elliptical galaxies in the dense clusters, something has indeed happened to skew the population in this way, so clusters are promising places to discover what processes might be involved. Fortunately, we can look back in time and see what kind of galaxies inhabited clusters when they were much younger and then surmise whether all galaxies were born spirals and then some were changed into ellipticals, or whether ellipticals have been around since the beginning (of galaxies). When we look out into space, we are looking back in time.

MAKING REAL CALCULATIONS OF HOW GALAXIES FORM

The ideas of galaxy formation outlined previously have been generated from years of thinking about the physics of gas and gravity, with some educated guesses about the conditions and processes under which stars might form. Although these notions underlie much of the thinking about galaxy formation, in truth, they did not extend our knowledge very far because galaxy formation is an example of the evolution of a complex system in which so many pieces are involved that the behavior is difficult to predict from simple analytical calculations. Perhaps you have heard about the field of study called *chaos* — the highly systematic but also highly complicated behavior of complex systems — the weather is my favorite example. Even if we could get rid of those troublesome butterflies in the Amazon, we would still find it hard to predict when it's going to rain in Indiana.

One change in the past two decades, growth in computer power, has helped us make progress in understanding the evolution of complex systems. Beginning with some very illuminating but still rough attempts to model the formation of galaxies in the 1970s, computers are now able in some ways to simulate what might have happened during the formation of galaxies. Why has such progress taken so long? The answer is that galaxy formation is a nightmare example of the "many body problem." Newton's laws of gravity can be used to figure out the motion of two or three bodies around each other, but if 10,000 bodies are all tugging on each other, about 100 million forces that will change continuously must be calculated. In fact, at least this many particles are needed to see the kind of complex behavior that real galaxies undergo as gravity assembles them. Perhaps we do not need a computer to follow the motion of every single one of the 100 billion stars in a forming galaxy — a staggering problem well beyond our reach — but we will need lots of calculating power if we are to follow accurately what happens when a cloud of gas begins to contract to form a galaxy.

How does such a calculation start? What are the initial conditions that the computer has to work with? We have had every reason to suspect, and we have now proven, that there were small ripples in density from place to place that could have grown into the structure we see today. The evidence has come from the exploration of **cosmic microwave background** radiation (see Chapter 1). This radiation provides us with a view of how smooth the Universe was about 300,000 years after the Big Bang. For example, if we look at a blank page of paper, we would get some idea of the kind of variations we're talking about — the page looks very uniform until we look closely. In fact, the variations in light intensity coming from a piece of apparently blank white paper are about 1%. The variations in light intensity expected from the microwave background were thought to be only about 1/100 of 1%. It took a long time and a lot of cleverness to push measurement techniques with telescopes in space to high enough accuracy to see such faint patterns. When no variations were found at the level of 1/100 of 1%, the level at which they were expected, astrophysicists were forced to conclude that the "ordinary" matter, often referred to as **baryonic matter** — protons, neutrons, and electrons — from which stars and people are made, could not have been chiefly responsible for the first stages of galaxy formation — the collection of matter by gravity. I believe this information was further compelling evidence for the existence of **dark matter**. As discussed in Chapters 1 and 7, this matter can be sensed from the influence of its gravity, but cannot be "seen," and it is probably the real driving force behind galaxy formation.

In a remarkable picture of the Universe in microwave light, which indicates how the Universe was 300,000 years after the Big Bang, we have now seen these faint patterns (see Figure 2.3). This experiment detected ripples at a level of 1/1000 of 1% variation from place to place. We are not yet sure where the **fluctuations in density** came from, although an interesting speculation is that they are unavoidable quantum fluctuations that were amplified enormously during a period of extraordinary growth in the size of the Universe, a period called the *inflationary epoch* (see Chapters 1 and 7). The pattern is not expected to be regular, like a checkerboard or herringbone, but more like the chaotic pattern of waves in the middle of the ocean — some large, more medium-sized, and many small. We even have a good guess for the relative proportions of different size waves in this early sea of matter — a form called $1/f$ that is very common in nature, for example, in the number of peaks of different sizes in a mountain range. In this type of distribution, how much of something there is varies inversely with the scale size; that is, something twice as large is twice as rare.

If one inputs a noisy pattern like this into a computer (as a very large set of numbers), with the dark matter of the Universe represented by millions of discrete mass points, the pattern prescribed will grow in contrast as time proceeds, according to Newton's laws of gravitation. Gravity will be strongest in the regions with a little higher density — even if only a little — and will work to increase the density there at the expense of the low-density regions. Unlike oceans, where waves crash and dissipate, here the waves grow and grow, which makes the contrast of the initial pattern, at first only 1/1000 of 1%, steadily increase, until in 1 billion years or so the variations become about 100%, which is the epoch at which we could say that galaxies are really forming. For the first time we can see the evolution of the complex matter distribution of millions of mass points with a faint pattern imposed — the way we imagine the Universe to have started — and we can see how the lacy tapestry of galaxies strewn through space might have developed. The complex behavior of the many-bodied gravitational system was too hard to predict with pencil and paper; we are having to reconstruct the Universe in a computer and let it evolve itself. As the computer simulations grow in complexity and power, they are surprising us with the reproduction of detail that we once thought was inexplicable. In a way, the entire Universe was a giant **N-body computer** running the "gravity code." We are just beginning to understand the consequences of treating this problem correctly.

The early failure to find fluctuations in the microwave background at the level of 1/100 of 1% has convinced many of us that dark matter is the driving force in making these galaxies. The idea that dark matter is the main component of mass in the Universe, along with other evidence, leads us to believe that each galaxy that we see is really a concentration of ordinary baryonic matter into stars floating in a considerably bigger pool of dark matter that does little more than just sit there and self-gravitate, sort of the ultimate couch potato. The substantial contrast between where the ordinary baryonic matter is and where it is not in today's Universe tells us that the true ripples in the density of matter were larger than the microwave light would lead us to believe — otherwise, there is not time to grow the faint patterns seen in the cosmic microwave background into today's galaxies and clusters. That we are misled is indeed possible, in fact, unavoidable, if there is a different kind of matter, a weakly interacting dark matter whose ripples in density have been growing from an even earlier time than we observe in the cosmic microwave background. The key point is that ordinary baryonic matter and light interact so strongly that they would have been locked together for the first 300,000 years; for this reason, the ripples in the ordinary matter could not at first grow

FIGURE 2.3

Two pictures of the sky in microwave (radio) light taken by the Cosmic Background Explorer (COBE) satellite. The *top* picture shows the smooth variation in intensity around the sky produced by the motion of the solar system at 370 km/sec relative to the average Hubble flow (see Chapter 1). With this gross variation removed, the more sensitive measurements of the *bottom* picture show a faint, blotchy pattern believed to be giant embryonic matter concentrations in the early Universe, concentrations that will grow into the structure we see today. (Courtesy of NASA-Goddard Space Flight Center and the COBE Science Working Group.)

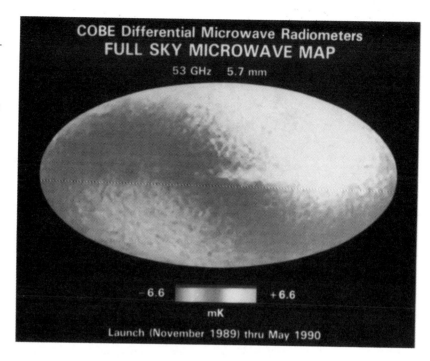

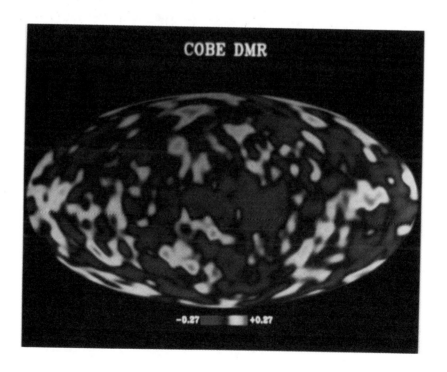

under the influence of gravity — the pressure of light itself prevented it until the Universe became "transparent" at an age of about 300,000 years. But the exotic dark matter was unaffected by the "slow growth ordinance" of the ordinary baryonic matter. It could have been well along in the process of making crevasses and pockets into which the protons, neutrons, and electrons would fall after their release from the sea of light. This would explain how the Universe could have evolved such great contrast by today, at an age of about 10 billion years, from what appear to be too-small ripples 300,000 years after the Big Bang. The ripples in the all-important dark matter would have been significantly larger than those that have been measured in the cosmic background radiation.

USING THE HUBBLE SPACE TELESCOPE AND N-BODY COMPUTERS TO GO FURTHER

Computer modeling procedures are telling us about how galaxy-size lumps might have formed, and something about how they are distributed in a typical volume of space (see Figure 2.4). These are very important results; we are now becoming confident that we can understand why there are galaxies and why they are distributed in space in lacy patterns and clusters. At this point, computers are following only the growth of the pattern in the dark matter. We are not yet making progress on how galaxies develop into different forms, for example, the difference between a spiral and an elliptical galaxy, for these differences pertain to the baryonic matter that makes stars. To address this issue, my group, as well as other astronomers, are pointing the Hubble Space Telescope (HST) at distant galaxies, galaxies so far away that the light has taken billions of years to reach us. This is a direct attempt to find out how and when different types of galaxies came to be — by looking back into history directly, we think we can actually watch the Universe as it accomplished all this, what one astronomer has called "traveling the time machine."

My group includes Gus Oemler and Ian Smail in the United States, Richard Ellis and Ray Sharples in the United Kingdom, Harvey Butcher in the Netherlands, and Warrick Couch in Australia. We have targeted clusters of galaxies, rich aggregations that have hundreds of galaxies in a volume scarcely larger than that encompassing the Milky Way and its neighbor Andromeda. We know that today such clusters of galaxies are replete with elliptical galaxies, but poor in spirals, which is elsewhere the most common type. What were clusters like 4 or 5 billion years ago, when the Universe was one-third younger than it is today?

The answer, from our HST images and hinted at by observations with giant telescopes on the ground for more than a decade, is quite startling — only 4 to 5 billion years ago many star-forming galaxies were in these clusters, galaxies most closely related to today's spirals. What happened to them? Why have they disappeared from clusters by the present day? Because we believe that these clusters are most certainly ancestors of the kind of clusters we see around us today, we are puzzled about why so many spiral galaxies have disappeared in such a relatively short time (by cosmological standards) — it is like looking at a picture of yourself when you were 30% younger and not recognizing yourself. We are tempted to regard this information as evidence that the "nurture" picture of galaxy formation is correct — perhaps all the spiral galaxies ran into each other and formed new elliptical galaxies, the ones we see in clusters today. Indeed, we see examples of what appear to be such "galactic wrecks" in these clusters. But my group is beginning to believe that this is not the whole story, perhaps

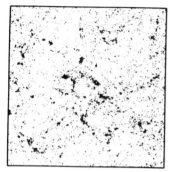
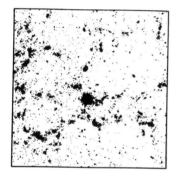

FIGURE 2.4

An N-body computer simulation showing the evolution of a smooth model Universe into a lumpy one through the influence of gravity. As time passes (*top to bottom*), the contrast of the pattern grows as gravity works to slow the expansion of the Universe in regions of high density. The expansion itself has been removed from the diagram; each subsequent picture has been zoomed out by the same factor that the model Universe has expanded, so that all three pictures contain exactly the same collection of matter. (From Davis et al., 1985. *Astrophysical Journal* 292: 371. Used with permission.)

not even the main story. For one thing, we see many ellipticals in these younger clusters, as if the ellipticals have been there all along. In itself, this seems to favor the "nature" picture of galaxy formation. Certain other kinds of observations tell us that the ellipticals were already old in terms of their stars and their structure, even 5 billion years ago. We might not expect such age if many of them had recently been made from the collision of a couple of spiral galaxies (see Figure 2.5).

As usual, our imaginations did not anticipate the inventiveness of nature herself. We have recently speculated that spiral galaxies in these clusters looked rather ragged, even when they were far from a collision with another galaxy. We wondered whether we were underestimating the effect of a spiral running the obstacle course of a rich cluster; although only a few galaxies might actually run into one another, we know that each galaxy will be pulled and tugged by the gravity of many neighbors as it goes whizzing through, and the entire cluster has a strong gravitational effect like the tides raised on Earth by the Moon. Thus, stars may be stripped from spiral galaxies and the all-important cloaks of dark matter that these galaxies are wrapped in may be sent into violent flapping or be completely wrested away — this could threaten a galaxy's stability and its survival (see Figure 2.6).

New computer simulations by George Lake, Neal Katz, and Ben Moore at the University of Washington are finding evidence for these kinds of processes, which are called *galaxy harassment*. Old paper and pencil approximations and even the previous computer simulations may have drastically underestimated these effects. The increase in computer power allows Lake and collaborators to add many more particles and make a much more realistic simulation — pundits in this field refer to the improved simulation as "greater dynamic range," and it does seem to make a difference. In the simulations we can see the drastic effect on a single galaxy as the gravitational effects of its neighbors and the cluster itself pull and jostle the galaxy as it speeds through the cluster at 1000 miles per second. This may be the principal reason that spirals do not survive in rich clusters — they are essentially shredded by their environments. If it is also true that the ellipticals are old and well established, even 5 billion years ago, we may have evidence that galaxy destinies were established early, but that the effects of environment were sometimes sufficient to tear apart the early work, especially in lower-density spiral galaxies.

These simulations are new and, as the authors admit, still imperfect. Huge amounts of time with supercomputers will be required to fully explore what is going on. For example, the simulation described above follows the harassment of a single galaxy but is not able to model the mutual harassment of a full population of clusters galaxies as they tear on each other simultaneously. This significant shortcoming will be rectified in the years ahead. Also, we can hope that future models will include a treatment of how new stars can form as the galaxies are pushed and pulled upon — rather than just following the dark matter — this is certainly a large part of the story of how a real galaxy behaves (see Figure 2.7).

Until the simulations become realistic enough to investigate all the processes involved, we must press further with the HST and with a new generation of giant telescopes on the ground to look for answers even further back in time. Already we are uncovering examples of what we think are galaxies at lookback times of 10 billion years, when the Universe was perhaps only 20% to 30% as old as it is today. The first observations provide tantalizing clues that galaxies were in messy bits and pieces at this early time, just in the process of coming together to make the spiral and elliptical galaxies we see today. There is the real hope that, in time, we will develop the tools and understanding to see galaxies like our own actually in the act of formation from the cooling gas of the early Universe (see Figure 2.8).

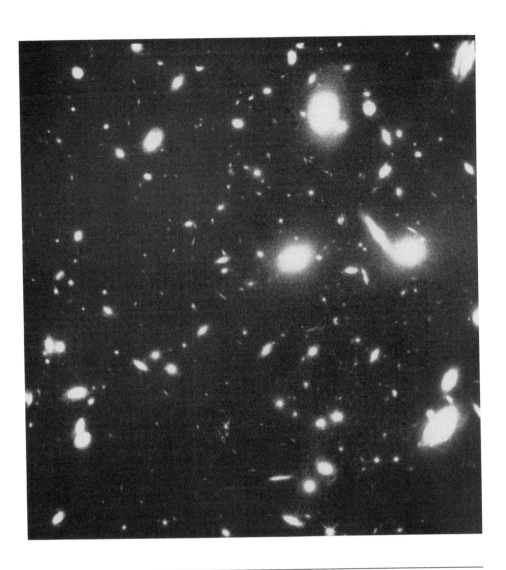

FIGURE 2.5A

A section of a rich cluster of galaxies imaged with the HST. All the extended images are galaxies as large or larger than the Milky Way — only very few stars from our own Milky Way galaxy show in the foreground. The cluster is so distant that the light travel time is about 4 billion years; therefore, this cluster is observed the way it appeared 4 billion years ago. At that time, apparently, the elliptical galaxies appeared much as they do today, but the spiral galaxies, nearly absent from rich clusters of galaxies today, were also abundant. A preliminary conclusion is that elliptical galaxies are unchanged since very early in the history of the Universe, but that spirals may undergo continuing evolution.

COMPLEXITY — UNDERSTANDING THE EVOLUTION OF THE UNIVERSE AND TURNING THE COPERNICAN REVOLUTION

Even with the monumental discoveries of our time, such as the extent and history of the Universe, the structure of the DNA molecule, and the quantum nature of matter,

FIGURE 2.5B

This is the second portion of the cluster shown in Figure 2.5A.

the greatest achievement of the scientific enterprise, to me, is the answer to the question "How did we get here?" which is captured in a single key concept, "complexity," and the realization that the evolution of the Universe, from the Big Bang to the present, has been a grand process in the building of great complexity from utter simplicity. If the answer to the question "What are we?" is "fantastically complex chemical machines (or organisms)," then understanding our relationship to the Universe involves recognizing that the Universe has been growing in complexity, engaging in the process of making possible the intricate chemistry of life and conditions in which it flourishes.

The first moments of the Big Bang, a screaming hot sea of massive particles and light, was as simple as anything has ever been — if we equate complexity with intelligence, there has never been anything since as stupid as the Universe of the Big Bang,

FIGURE 2.6

Examples of interacting or merging galaxies in the cluster shown in Figure 2.5. *Top*: Two galaxies (each shown at two contrast levels) that are probably the result of the merger of two galaxies. *Bottom*: Glancing encounters between galaxies that have pulled stars and dust from each other and have spewed the material into intergalactic space.

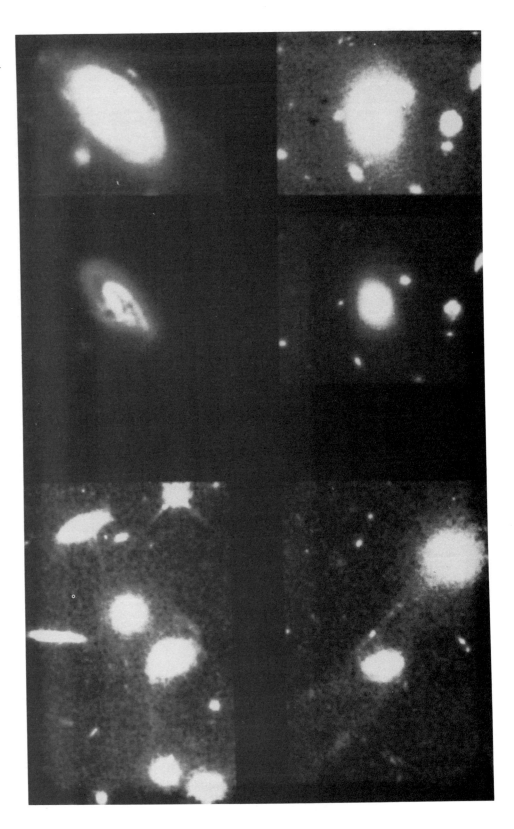

FIGURE 2.7

A montage of spiral and peculiar galaxies in the cluster of Figure 2.5. *Top six rows*: The best examples of spiral galaxies in the cluster; it is apparent that few, if any, show the regular spiral patterns we are accustomed to seeing in today's Universe (see Figure 2.2), which suggests that the earlier spirals were either more juvenile and thus less well-assembled or that they have been jostled by encounters with other galaxies in the cluster. *Bottom two rows*: Examples of peculiar galaxies that are thought to be the result of mergers or strong encounters with other galaxies.

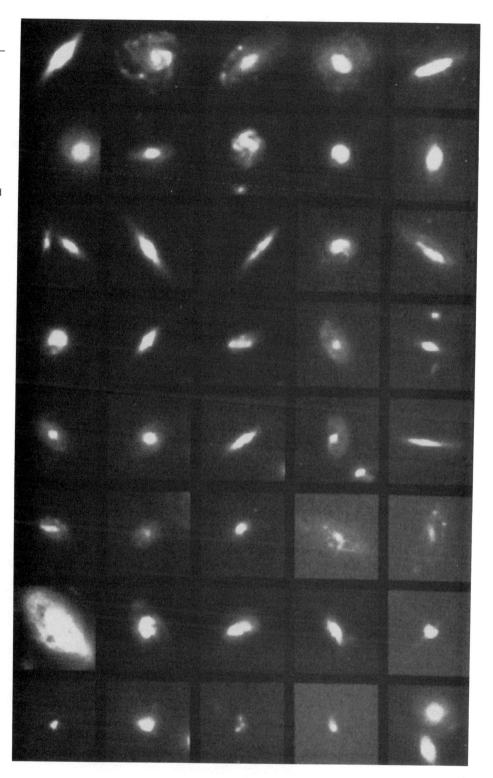

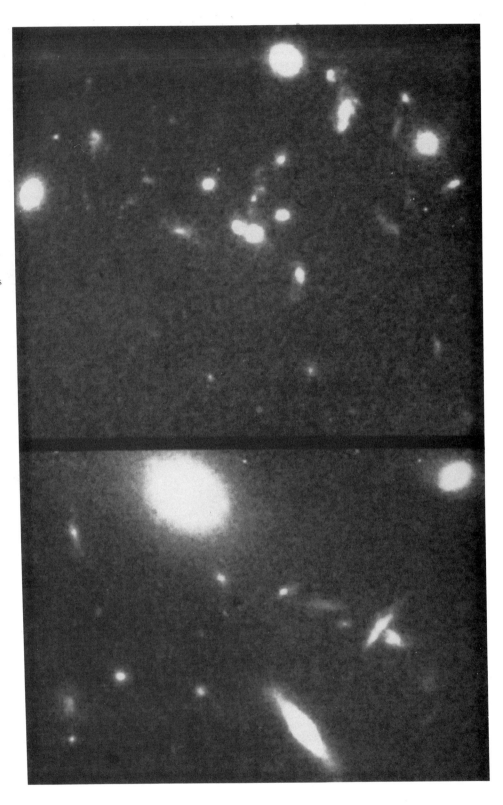

FIGURE 2.8

A view even further into the past? *Top*: One object is a quasar, a bright beacon at the center of a galaxy seen long ago, approximately 10 billion years in the past. The quasar is the brightest of the objects at the *center* of the panel. The other faint blobs near this quasar may be companions of the quasar. If so, they are some of the earliest examples of what galaxies may have been like in the very young Universe. Continuing the trend seen at a lookback of 4 billion years, the very young galaxies seem to be very ragged — there is little similarity to the normal Hubble types seen today (see Figure 2.2). *Bottom*: Another HST image of galaxies in the field near the distant quasar, just to the right of the top panel.

and there never will be. The behavior of all the matter and energy was governed, we believe, by gravity and by a superforce that allowed continued variation from one manifestation of energy to another but did not allow the slightest pattern to endure. This was a perfect symmetry, looking the same at all space and at all times. The evolution of the Universe from the Big Bang, the state of ultimate dullness to the marvelous, complex world of today, is a story of broken symmetries, of the Universe expanding and cooling and, as it did so, breaking the boring symmetries. As discussed further in Chapter 7, we might compare this process to the way the symmetry of liquid water, which knows no direction or macroscopic structure, breaks to form ice, which is much more interesting because it has great diversity of structure, or snowflakes — we all remember hearing as children that no two snowflakes are alike. In an analogous way, the superforce broke into the strong nuclear force, the weak force, and the electromagnetic force, which allowed interesting complexity to build where only simplicity had been before.

Early in its existence, the Universe had been made up of 75% hydrogen (pretty simple stuff) and 25% helium (pretty useless stuff for chemistry), and none of the crucial elements that are needed to make anything interesting, like a rock or an enzyme. Later, in the very centers of stars, the strong and weak force became the agents for building the heavier nuclei of carbon, nitrogen, oxygen, iron, and the like (see Chapter 5). The new wide variety of building blocks made the Universe more interesting. And the electromagnetic force, a broken symmetry of the once monotonous superforce, was now available to link together these nuclei at the atomic level — voila, chemistry. The development of chemistry left the Universe with fantastic gains in possible states: complexity.

But where could the gain in complexity be realized, to be able to accomplish something? Here is where gravity, the fourth of the fundamental forces in our Universe of broken symmetry, comes into play. In the primeval Universe, gravity built the patterns by first gathering energy and then matter into subtle waves, then patterns, and finally structures such as galaxies and their stars. Without gravity, the Universe would have stayed smooth and boring; with gravity it built stars where the strong and weak force could build atoms and planets where the electromagnetic force could put them together. The complex chemistry that made life could not happen just anywhere — stars are too hot and the space between them is too thin. Planets have a delicate balance of temperatures and densities favorable for chemistry with fantastic variation, the hallmark of life (see Chapter 6).

If the Universe had not, at several different stages in its evolution, broken symmetries, which allowed the building of increasingly complex structures, then the story would be over by now. In particular, the last step, the development of biology, which may have already happened throughout the Universe, is a step so meaningful that it has changed everything, for it has released the future of the Universe from a fate solely at the hand of physics. Without life, the future of the Universe would be a sad tale of devolution to cold, dead galaxies with abundant black holes, perhaps the decay of matter itself.

It is true in one sense that the Universe is dying. Entropy is increasing — there are more disordered states than ordered ones. Overall, the Universe is moving inexorably toward what our colleagues in the nineteenth century called "the heat death" — the Universe grows thinner and colder and the nuclear fires that built the chemical elements begin to run out of fuel. It would seem that nothing but cold, still death awaits everything. But, strangely, the Universe has beaten the rap, by developing cool, not cold, places in which complexity grows at an exponential pace. On a global scale, the battle against entropy will eventually be lost, but we and other creatures like us inhabit rising castles that tower over the doomed landscape. This is the legacy of the process of which we have discovered ourselves to be a part.

The presence of life changes everything. Considering what has happened on Earth, can we even imagine what effects life might have on the evolution of the Universe in another billion years? What changes might be in store that textbook physics could not predict? It seems to me there is much more in the Universe's future than a cold or fiery death — the rise of beings promises to provide the real changes for the Universe to come.

This is what brings us back to the center of the Universe. It is often said that the message of modern astronomy has unintentionally convinced people that humans are small and insignificant in relation to the Universe. I urge all of us to reconsider this doctrine and to spread the word that the true message of astronomy is that complexity and intelligence is the true gift, and in that respect we are very much at one of the centers of our Universe. A passage from my book, *Voyage to the* **Great Attractor**, expresses how I feel about this important issue.

> We continue to take the wrong lesson from what we are now learning. An astronaut who had taken a tethered spacewalk while on a Gemini flight was recently asked whether the experience had changed him. He had been struck, he recalled, by how small and insignificant were the Earth and the human adventure, "like an ant crawling across the Sahara desert." Exactly. The ant, astronomically outnumbered by the grains of sand, overwhelmed by the size of the inhospitable desert, is nevertheless the greater marvel, by far.
>
> It is time to take full stock of the discovery that life is the most complex thing we know of in the Universe, and, as such, most worthy of our admiration. Yes, the Universe dwarfs our world in size and immense power. But the Universe of stars, galaxies, and vast gulfs of space is so very, very simple compared to us and our brethren life forms. If we could but learn to look at the Universe with eyes that are blind to power and size, but keen for subtlety and complexity, then our world would outshine a galaxy of stars. Indeed, we should marvel at the Universe for its majesty, but we must truly be in awe of its greatest achievement — life. The Universe has invented a way to know itself, a way to explore itself, to propagate subtle and intricate design throughout itself. The process of creation has been coming in our direction for more than 10 billion years, and now, with us and what others there may be like us, the flow is turned back to the Universe, from whence it came. We are ready to explore the Universe; whether we launch only our senses and our minds or send our bodies as well, we are bound now to take our gift out into the Universe, perhaps to change it forever. We should feel at the center of our Universe for, in a very real sense, we are its point.

FURTHER READING

Dressler, A. 1991. Observing galaxies through time. *Sky & Telescope 81*: 126.

Dressler, A. 1993. Galaxies far away and long ago. *Sky & Telescope 85*: 22.

Dressler, A. 1994. *Voyage to the Great Attractor* (New York: Knopf).

Ferris, T. 1982. *Galaxies* (New York: Steward, Tabori, & Chang).

Lightman, A. 1991. *Ancient Light* (Cambridge: Harvard University Press).

Silk, J. 1994. *A Short History of the Universe* (New York: Scientific American Library).

Time-Life Books (Eds.). 1988. *Voyage Through the Universe: Galaxies* (Richmond, VA: Time-Life Books).

CHAPTER 3

THE ORIGIN OF STARS AND PLANETS

Fred C. Adams

INTRODUCTION

The formation of stars and planets is one of the most fundamental problems in astrophysics. In recent years, a lot of progress in this area has been made. In particular, we now have a fairly successful paradigm that provides the cornerstone of our current understanding of the star formation process. Within this paradigm, the agreement between observations and theory is quite good, especially for the case of low-mass stars.

The process of star formation basically boils down to a war between entropy and gravity. The same is true for stellar evolution in general. Here, "entropy" manifests itself as pressure forces and turbulence. Loosely speaking, gravitational forces tend to pull things together, whereas entropy tends to spread things out. As this chapter illustrates, this war between gravity and entropy takes place on many different scales of size and mass. Furthermore, this war largely determines how stars form and evolve.

The chapter begins with an overview of the current theory of star formation. Next, circumstellar disks, which play an important role in the formation of stars, are presented. Here, several mechanisms that have been proposed to determine the dynamical evolution of disks are considered; these mechanisms include gravitational instabilities and viscous effects. Circumstellar disks also provide the setting for planet formation, which is also considered. A discussion of **initial mass function** (IMF), the distribution of stellar masses at their births, follows. A theory of initial mass function is essential for understanding the effects of star formation on galactic evolution, galaxy formation, and other important astrophysical topics; however, such a theory remains a fundamental unresolved issue. The chapter concludes with a summary and a discussion of other unresolved issues.

STAR FORMATION IN MOLECULAR CLOUDS

In the star-formation paradigm that has emerged in the past decade (Shu, Adams, and Lizano, 1987; Lada and Shu, 1990), stars form within **molecular clouds**, which are large, massive aggregations of molecular gas. These clouds are much denser and much colder than the surrounding interstellar gas; they have typical number densities n of approximately 100 cm^{-3} and typical temperatures T in the range approximately 10 to 35 kelvins (K) (in ordinary air at room temperature n is about 10^{19} cm^{-3} and T is about 293 K). In fact, stars are forming today in nearby molecular clouds, which provide a useful laboratory to study the star-formation process (see Color Plate 3.1). Molecular clouds are much larger than stars and typically have masses 10^4 to 10^6 times the mass of the Sun ($M_\odot$).

Stars actually form out of the collapse of **cores** of molecular clouds. The core regions, small subcondensations within the much larger molecular clouds, are supported, in part, by magnetic fields. Such fields provide a source of pressure which helps support the cores against gravitational collapse. The cores evolve through a process called **ambipolar diffusion**, in which the magnetic fields slowly diffuse outward and the inner regions of the core become increasingly centrally concentrated. The magnetic contribution to the pressure support decreases with time until thermal pressure alone supports the core against its self-gravity (at least in the central regions). At this point, the core is in an unstable equilibrium state, which represents the initial conditions for the subsequent dynamic collapse.

When a core undergoes collapse, a small pressure-supported object (i.e., the forming star itself) forms at the center of the collapse flow. The cloud cores are rotating and hence contain a substantial amount of angular momentum (i.e., spin). The infalling material with higher specific angular momentum collects around the forming star and creates an accompanying **circumstellar disk**. The presence of this disk is thus a natural consequence of the law of conservation of angular momentum. This phase of evolution (often denoted as the protostellar phase) is thus characterized by a central star and disk, surrounded by a flow of gas and dust falling inwards toward the central object. The characteristics of this infalling envelope largely determine the characteristics of the radiation that is emitted by the object during this phase of evolution; radiation is then detected by astronomers at far away places such as Earth.

As a **protostar** evolves, both its mass and luminosity (rate of energy output) increase. The protostar eventually develops a strong stellar wind that breaks out through the infall at the rotational poles of the system and creates a bipolar outflow. This phase of evolution is denoted as the **bipolar outflow** phase. For much of this phase, the outflow is very narrow in angular extent and infall takes place over most of the solid angle centered on the star. The outflow gradually widens in angular extent and the **column density** of the infalling material gradually decreases. In other words, as time goes on, the star becomes less deeply embedded in the molecular cloud core.

The outflow eventually separates the newly formed star/disk system from its parental core and the object becomes a young star. The latter stage of evolution is denoted as the **T Tauri** phase or the **pre-main-sequence** phase. During this stage of evolution, the system often retains its circumstellar disk. Planets might form within the disk during this evolutionary epoch. Although the newly created star itself is optically visible, it does not have the right internal configuration to generate energy through nuclear fusion of hydrogen. Instead, the star generates most of its energy through gravitational contraction. As the star contracts, however, the central temperature increases until hydrogen fusion can take place. When the ignition of hydrogen occurs, the star is fully formed. The paradigm of star formation can be neatly packaged in terms of four stages, as shown in Figure 3.2.

Molecular Clouds — The Birthplace of Stars

Molecular clouds are the birthplaces of stars. One important aspect of these clouds is that they are not collapsing as a whole (Zuckerman and Palmer, 1974). This finding is somewhat surprising because the total amount of self-gravity in these systems is generally much larger than the amount of thermal pressure support; that is, the gravitational force within the cloud is generally stronger than the ordinary thermal pressure force (see Shu, Adams, and Lizano, 1987). As a result, the clouds must be supported by some nonthermal means. The most viable candidates for providing this support are magnetic fields and "turbulence."

Magnetic fields provide an important additional source of support and can prevent molecular clouds from collapsing. Very roughly, for a cloud of a given size and mass, a minimum magnetic field strength B is required to prevent collapse (Mouschovias and Spitzer, 1976). Observations of magnetic fields in molecular clouds show that the magnetic field strengths are generally large enough to support the clouds against gravitational collapse (Myers and Goodman, 1988; Heiles et al., 1993). In particular, the field strengths are typically in the range 10 to 30 μG (these field strengths are about 10^5 times smaller than the strength of the magnetic field of Earth but the fields in the clouds cover an enormously larger volume of space). In addition, we can determine the geometry of the magnetic fields in these clouds. The polarization vectors of background starlight seen through the clouds are thought to trace patterns in the magnetic field lines. Observations show that the polarization vectors are very uniform and well ordered (Goodman, 1990). Thus, we infer that the magnetic field itself also has well-ordered structure. In other words, the magnetic field is not extremely tangled; it retains a more-or-less uniform component.

Turbulence in molecular clouds has begun to be studied, but definitive results are difficult to obtain (Scalo, 1987). One important result is that the observed linewidths of molecular transitions often contain a substantial nonthermal contribution. In this case, nonthermal motions of parcels of gas cause the widths of molecular transition lines to broaden as a result of the Doppler effect; that is, the relative motions of the gas cause the frequency of the emitted light to shift and create apparently broader lines. These observations thus imply that the cloud fluid has substantial motions on small-sized scales and that these motions are supersonic in character; in other words, the cloud exhibits "turbulent" behavior.

The formation of substructure within molecular clouds remains an important problem. Molecular clouds have highly complicated substructure on virtually all size scales that one can observe (Houlahan and Scalo, 1992; Wood, Myers, and Daugherty, 1994; Wiseman and Adams, 1994) (see Figure 3.3). In particular, these clouds do not have a single, well-defined, characteristic density; instead, the clouds have a very wide range of different densities. The geometrical distribution of the cloud material is also quite complicated. Cloud structures have been described in terms of "sheets," "filaments," and quasispherical clumps of cloud material (Blitz, 1993). At present, no definitive theory exists for the formation of substructure within molecular clouds. However, we have

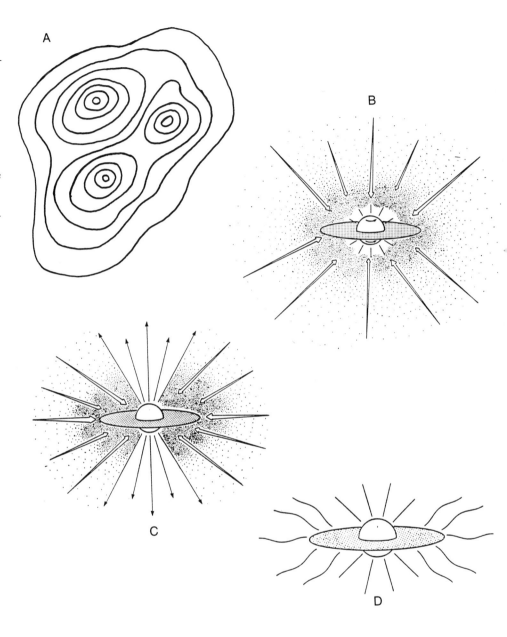

FIGURE 3.2

The four stages of star formation. (a) Molecular cloud cores form within molecular clouds. (b) The cores collapse to form protostars. (c) Bipolar outflows break through the infall along the rotational poles of the system. (d) The young stars and their circumstellar disks become optically revealed. Lines represent visible photons emitted by the star and longer wavelength infrared photons emitted by the disk.

one clue to this puzzle. Whenever a cloud begins to collapse, a wide spectrum of wave motions can be excited (Arons and Max, 1975), and these wave motions may provide part of the explanation for the observed clumpy structure (Figure 3.3).

Core Formation Through Ambipolar Diffusion

Molecular cloud cores represent the cloud structure on small scales of size and mass. These scales are small compared with the overall molecular cloud, but are still much larger than the scales of the forming stars themselves. The formation of the cores occurs through the process of ambipolar diffusion (Mouschovias, 1976 and 1978;

FIGURE 3.3

Column density map of the Taurus molecular cloud as produced from 100 μm dust emission maps made by the NASA infrared astronomical satellite (IRAS). (From Houlahan and Scalo, 1992; Wiseman and Adams, 1992.)

Shu, 1983; Nakano, 1984; Lizano and Shu, 1989). This process occurs as follows: The cloud is very lightly ionized; only about one particle per million carries a charge. The charged particles — the ions — interact directly with the magnetic field, but the neutral particles (which make up most of the mass) do not. (Magnetic fields can only exert forces on charged particles.) The only interaction that the neutral particles have with the magnetic field is through interactions with the ions. These interactions occur through a frictional force that, in turn, depends on the relative motion between the ions and the neutrals. Thus, the ions can exert a force on the neutrals only if the two species are moving with respect to each other. The important physical implication of this situation is that the ions (and, hence, the magnetic field) must be slowly moving outward relative to the neutral component, which is itself slowly contracting because of gravity.

The diffusion process takes place on a time scale that is generally much longer than the free-fall collapse time scale of the cloud (the time it takes the cloud to collapse in the absence of any pressure forces). In fact, the magnetic diffusion time scale is roughly 10 times longer than the dynamic free-fall time scale for typical molecular clouds. As a result, the process of losing magnetic flux (i.e., magnetic pressure support) provides the bottleneck in the star formation process and keeps the efficiency of star formation fairly low ($\sim 1\%$) in molecular clouds. In other words, only a small fraction of the molecular cloud material ends up as stars.

The mass scale defined by the molecular cloud cores is still much larger than that of the forming stars, by a factor of approximately 100. Hence, the interstellar medium cannot (by itself) determine the mass scale of forming stars. Instead, the core regions provide the initial conditions for star formation; in particular, they provide the starting conditions for protostellar collapse, as described in the following section.

Molecular Cloud Cores — Initial Conditions for Collapse

Stars form within molecular cloud cores, and the observed core properties provide the initial conditions for protostellar collapse. In nearby molecular cloud complexes that are forming low-mass stars, the cores are observed to be slowly rotating and have a single constant temperature. In our current (idealized) theory of star formation, the cloud cores (the initial state) can be described by two physical variables: the temperature and the (uniform) rotation rate of the core. Current observations of cloud cores indicate temperatures in the range 10 to 35 K (Myers, 1985). The measured rotation rates of these cores are extremely slow — the measured angular velocity is about 3×10^{-14} rad s^{-1}, which means that it takes about 10 million years for the core to make a single complete rotation. At small-sized scales, the cloud cores are highly centrally concentrated and are expected to have density profiles in which the thermal pressure force exactly balances the gravitational force at all radii (Shu, 1977; Terebey, Shu, and Cassen, 1984). The central parts of the core thus have much higher densities than do the outer parts.

Recent observational work (Myers and Fuller, 1992) indicates that cores of higher mass have a slightly different structure than that of the simple model described above. For sufficiently large-sized scales ($r \sim 1$ pc) and low density ($n < 10^4$ cm^{-3}), observations suggest that the cores have substantial nonthermal motions with random velocities, v, that vary with density, ρ, according to the relation $v \propto \rho^{-1/2}$. If this velocity is interpreted as a transport speed, then a turbulent or nonthermal component to the pressure can be derived (Lizano and Shu, 1989). In other words, the total pressure includes a thermal and a nonthermal component. Both must be considered for high-mass cores.

Protostellar Collapse

Protostars are objects still gaining mass through infall. For low-mass protostars, that is, forming stars with masses comparable to that of the sun, the radiation field is not strong enough to affect the infalling envelope and the dynamic collapse is thus decoupled from the radiation. Molecular cloud cores tend to collapse from inside out (Shu, 1977); in other words, the central part of the core collapses first and successive outer layers follow. The collapse scenario naturally produces a core/envelope structure — a pressure-supported object (the forming star itself) lies at the center and is surrounded by an infalling envelope of dust and gas. The inside-out collapse progresses as an expansion wave propagates outward at the sound speed. Outside the location of the expansion wave, the cloud remains static and has no information that collapse is taking place in the interior. Inside the expansion wave radius, the material falls inward and approaches free-fall velocities. The density distribution of this infalling envelope of dust and gas is nearly spherical outside a centrifugal radius, which we denote as R_C. The centrifugal radius provides an important length scale in the protostellar envelope; it represents the position where the infalling material with the highest specific angular momentum encounters a centrifugal barrier (caused by conservation of angular momentum) (Cassen and Moosman, 1981; Terebey, Shu, and Cassen, 1984). Inside the centrifugal radius R_C, the collapse flow is highly nonspherical as particles spiral inward on nearly pressure-free trajectories. In the region immediately surrounding the star, the high temperatures evaporate the dust grains and all the material is gaseous (Stahler, Shu, and Taam, 1980).

There is no mass scale in this collapse scenario. Instead, the collapse flow feeds material onto the central star and disk at a well-defined mass infall rate $\dot{M}$. In the simplest case of an isothermal cloud core, the mass infall rate is a constant in time and depends only on the initial temperature of the molecular cloud core (Shu, 1977); when additional sources of pressure are taken into account, the mass infall rate increases with time. Typical mass infall rates lie in the range of 10^{-6} to 10^{-5} $M_\odot$ yr^{-1}; in other words, the time scale for forming a star like the Sun is 10^5 to 10^6 years. The formation time is extremely small compared with the lifetime of a Sun-like star and the age of the Universe (both these time scales are $\sim 10^{10}$ yr).

Protostellar Radiation

The radiation emitted by forming stars (protostars) is important because it allows for a test of the underlying theory and for a means of identifying protostellar candidates. The radiation field for protostellar objects can be divided into three separate components: the direct radiation field from the star, the direct radiation field from the disk, and the diffuse radiation field emitted by the infalling dust envelope. Both the star and disk can actively generate energy and thus emit radiation; like our Sun, the star and disk emit most of their radiation at optically visible wavelengths, although the disk also emits a substantial amount of energy at infrared wavelengths. However, much of the star/disk radiation is highly attenuated by the infalling envelope; in other words, most of the luminosity is absorbed by dust grains in the envelope and then re-radiated at longer (far-infrared) wavelengths that can escape more easily from the region. Hence, the spectrum of radiation that we actually see is determined mostly by the properties of the infalling envelope and is largely independent of the spectrum of the initial star/disk system.

The protostellar **luminosity** has several different contributions, although the ultimate source of energy is gravity. At this early stage of stellar evolution, nuclear fusion of hydrogen is not yet taking place in the star. As material falls toward the central star/disk system, gravitational potential energy is converted to kinetic energy, which is then converted into radiation in several different ways. The infalling material produces shocks on the surfaces of both the star and disk and thereby dissipates energy. Additional energy is dissipated as the infalling interstellar material becomes adjusted to stellar and disk conditions. In particular, the infalling material does not have the same rotation speeds as the material already in the star and disk; the newly added material must release energy as it adjusts to the local conditions. In addition, disk accretion produces a substantial amount of luminosity; in this process, material in the circumstellar disk dissipates energy and transfers angular momentum outward. The main uncertainty in this picture is the disk accretion mechanism; however, disk stability considerations place fairly tight constraints on the allowed disk accretion activity. The total luminosity of the system is thus reasonably well determined and is a substantial fraction of the total available luminosity

$$L_0 = \frac{GM\dot{M}}{R_*} \tag{1}$$

where R_* is the stellar radius (which is typically a few times the radius of the Sun) (Stahler, Shu, and Taam, 1980) and M is the stellar mass. The luminosity L_0 given by Equation 1 represents material falling through a gravitational potential well (of depth GM/R_*) at the mass infall rate $\dot{M}$. The actual luminosity of the system is generally less than this maximum value because some material does not fall all the way to the stellar surface but remains in orbit about the star at larger radii. Thus, some energy is stored in the form of rotational kinetic energy and gravitational potential energy.

The diffuse radiation field of the infalling dust envelope can be determined through a self-consistent radiative transfer calculation (Adams and Shu, 1986; Kenyon, Calvet, and Hartmann, 1993). Such a calculation keeps track of all photons (the particles that make up the radiation field) as they travel outward through the envelope and become absorbed by the dust grains and re-radiated at longer wavelengths. The theoretical **spectral energy distributions** calculated from this protostellar model are in reasonably good agreement with observed spectra of protostellar candidates (see Figure 3.4) (Adams, Lada, and Shu, 1987; Myers et al., 1987). The spectral energy distributions show how much energy is emitted by the protostar at various wavelengths (or, equivalently, frequencies). These spectra generally have maxima at wavelengths of 60 to 100 μm. Almost all the energy is emitted at wavelengths much longer than that of visible light; this property of the sources explains why protostellar candidates have been identified only recently. The spectra of sources in the bipolar outflow stage of evolution are also well described by protostellar infall models; this result suggests that both infall and outflow are taking place simultaneously in such objects.

Because the density distribution of the protostellar envelope is known from the dynamic collapse solution and the temperature distribution is known from radiative transfer calculations, the spatial distribution of emission from protostars can also be calculated theoretically. Recent observations have produced spatial emission maps of protostars at millimeter and submillimeter wavelengths in the continuum. The measurements show that protostellar emission is extended and that the observed spatial profiles of emission are roughly consistent with the current theory (see Figure 3.5) (Walker, Adams, and Lada, 1990; Butner et al., 1991; Ladd et al., 1991).

The L–A_V Diagram for Protostars

Astronomers have traditionally studied and described stellar evolution by making use of the Hertzsprung-Russell (HR) diagram, which plots the luminosity of a star on the vertical axis and the surface temperature of the star on the horizontal axis. These two physical variables (luminosity and surface temperature) are used because they adequately characterize a given star and because they can be determined from observations. Ordinary hydrogen-burning stars live on a well-defined locus in this diagram known as the **main-sequence**; different points along the main-sequence correspond to stars of different masses. Once stars have used up the hydrogen in their cores, they evolve into new stellar configurations; as they evolve, they follow well-defined paths (usually called tracks) in the HR diagram. Thus, the HR diagram provides a useful tool with which to study stellar evolution.

Protostars cannot be placed in the HR diagram because they have no well-defined surface temperature — protostars have extended atmospheres (their infalling envelopes) and hence have spectral energy distributions that are made up of emission from many different regions of different temperatures. As a result, another approach must be invoked to diagrammatically represent the early phase of stellar evolution. For protostars, the total system luminosity L and the column density of the infalling envelope are the two most appropriate physical quantities that describe the system. The *luminosity* represents the rate of energy production by the object; the column density is

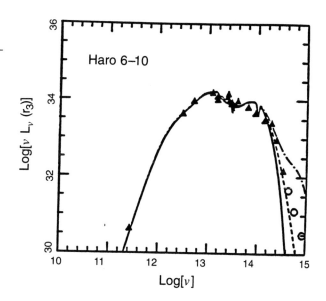

FIGURE 3.4

Observed and theoretical spectral energy distributions of protostellar candidate Haro 6–10. The frequency ν is plotted on the horizontal axis; the quantity plotted on the vertical axis represents the amount of energy emitted by the object at a given frequency of light. The curves represent theoretical models which include different treatments of the scattering of photons at near-infrared wavelengths. The symbols represent observational data. (Adapted from Adams, Lada, and Shu, 1987.)

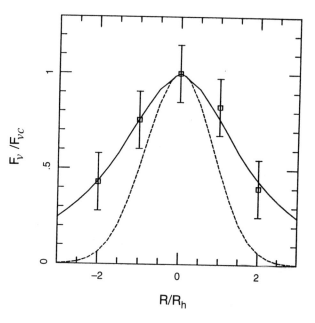

FIGURE 3.5

Spatial distribution of emission for protostellar candidate L1551-IRS5. In this figure, the amount of emission (vertical axis) is plotted as a function of the projected distance from the source center (horizontal axis). The solid curve is a theoretical model and the symbols represent observational data. The dashed curve indicates how a point-like (unextended) object would appear. Thus, the plot shows that this protostar is an extended object. (Adapted from Walker, Adams, and Lada, 1990.)

a measure of the thickness of the infalling protostellar envelope. We generally convert the column density into a quantity called the visual extinction A_V (very roughly, the *visual extinction* is the average number of times a photon of visible light will scatter as it travels through the envelope). We can thus use an L–A_V diagram (see Figure 3.6) to characterize the protostellar phase of evolution. The total system luminosity L_{bol} constitutes the vertical axis; the visual extinction A_V of the infalling envelope constitutes the horizontal axis (plotted backwards in keeping with astronomical tradition). For a given set of initial conditions, a protostar will trace out a well-defined track in the diagram as time proceeds (see Figure 3.6) (Adams, 1990). As time proceeds, protostars move from

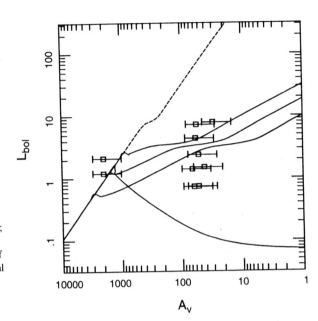

FIGURE 3.6

The L–A_V diagram for protostars in the Taurus molecular cloud. The total luminosity, denoted here as L_{bol}, is plotted on the vertical axis; the visual extinction A_V is plotted backwards on the horizontal axis. The solid lines in the diagram are the set of tracks calculated for varying core rotation rates; protostars move from the lower left toward the upper right as time proceeds. Open symbols represent observed sources; the somewhat large error bars in the horizontal direction arise because of the difficulty in measuring the visual extinction. (Adapted from Adams, 1990.)

the lower left toward the upper right in the L–A_V diagram. The visual extinction A_V decreases with time until the object moves off the right-hand side of the diagram and becomes optically visible. The newly formed star will then appear on the birthline in the HR diagram (Stahler, 1983; Shu, 1985; Stahler, 1988; Palla and Stahler, 1990). At this point in time, the object enters the T Tauri evolutionary phase.

The lowermost curve in Figure 3.6 shows the evolution of a protostar if accretion of disk material onto the protostar does not occur. In this case, all the material that falls directly onto the disk simply stays there at large radii and does not make its way onto the star. As time goes on, infalling material falls through an increasingly shallow potential well and, hence, the luminosity of the system decreases to values much lower than those of observed sources. Thus, some disk accretion must occur to account for the observed protostellar luminosities. If no disk accretion occurs, most of the infalling mass remains in the disk and does not become part of the star. Disk accretion must occur to account for the masses of forming stars (see later discussion of the initial mass function).

The Protostellar to Stellar Transition

The transition from an embedded protostellar object to an optically revealed young star may be accomplished through the action of a powerful stellar wind. The presence of such winds (and related outflow phenomena) was first established observationally (Lada, 1985). A presently favored theoretical description of these phenomena is a centrifugally driven magnetic wind model (see Figure 3.7) (Shu et al., 1988; Shu et al., 1994). This mechanism can be described very roughly as follows: The star generates a relatively strong magnetic field that rotates with the star. The field strength must be about 1000 G, which is large compared with the value of about one G for the Sun today. Some field lines are open; that is, they do not attach back onto the star, but continue out to spatial infinity. Parcels of ionized gas flow along these field lines (but not

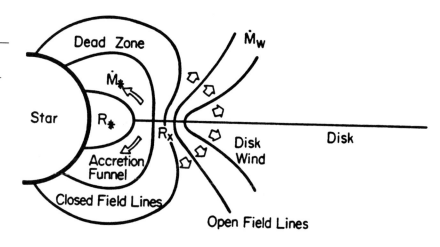

FIGURE 3.7

Schematic diagram of the centrifugally driven magnetic wind mechanism. (Adapted from Shu et al., 1994.)

across the field lines). Thus, parcels of gas travel along the magnetic field lines in much the same way that beads can travel along a wire. In this case, however, the magnetic field lines are rotating so that the parcels of gas are flung outwards by the centrifugal force. These outwardly moving parcels of gas become the wind.

When young pre-main-sequence stars (often T Tauri stars) first appear in the HR diagram, they seem to appear on a well-defined locus, denoted as the *stellar birthline* (see Figure 3.8) (Stahler, 1983). The location of the birthline corresponds to stellar configurations that are capable of burning deuterium (Shu, 1985; Stahler, 1988). Deuterium is the most easily fused of all available nuclei; deuterium-burning in stars can take place at a central stellar temperature T_C of approximately 10^6 K, but a much higher temperature, T_C about 10^7 K, is required for the fusion of ordinary hydrogen. In any case, stars appear on the HR diagram as visible objects with just the right properties to burn deuterium. One explanation for this finding is that deuterium-burning produces stellar **convection**; in other words, the heat generated by deuterium-burning is carried toward the surface of the star by motions in the stellar fluid itself. These convective motions, in conjunction with differential rotation in the star, can amplify magnetic fields to fairly large strengths (100–1000 G). Because such magnetic field strengths are required for the centrifugally driven wind, the entire picture of stars determining their own mass through the production of winds is consistent.

Pre-Main-Sequence Stars and Circumstellar Disks

Ordinary stars have spectral energy distributions that have the basic form of a blackbody. However, young stars are often observed to have an additional infrared component to their spectra (Rydgren and Zak, 1987; Rucinski, 1985; Appenzellar and Mundt, 1989). The most likely explanation for the infrared excess is the presence of circumstellar disks. These disks can either be active or passive. *Active disks* generate their own energy through the process of disk accretion; *passive disks* have no intrinsic energy source and only absorb and re-radiate stellar photons. In either case, the disk radiates photons at infrared wavelengths and produces the observed excess emission.

In systems with passive disks, all luminosity is generated by the star itself. The disk is generally spatially thin and optically thick; in other words, the disk absorbs all the

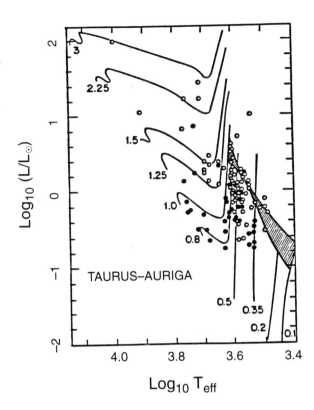

FIGURE 3.8

An HR diagram showing the stellar birthline. The stellar luminosity makes up the vertical axis; the stellar surface temperature is plotted backwards on the horizontal axis. The symbols represent the population of young stars in the Taurus-Auriga molecular cloud. Notice that almost all the stars lie below the hatched birthline. (Adapted from Stahler, 1988.)

light that is incident on it. When the inner radius of the disk extends down to the stellar surface and the outer disk radius is large compared with that of the stellar radius, the disk intercepts and re-radiates 25% of the stellar luminosity. The disk thus radiates as if it had its own effective luminosity (at most one-fourth that of the star) and adds excess infrared radiation to the spectral energy distribution of the system. The passive disk models produce the correct infrared excess for some observed T Tauri systems (Adams, Lada, and Shu, 1987; Kenyon and Hartmann, 1987).

The class of star/disk systems with active disks have appreciable intrinsic disk luminosity in addition to intercepted and reprocessed energy from the star. Through fitting their spectral energy distributions, we can estimate the basic physical properties of active disk systems (see Figure 3.9) (Kenyon and Hartmann, 1987; Adams, Lada, and Shu, 1988; Beckwith et al., 1990; Adams, Emerson, and Fuller, 1990; Calvet et al., 1994). In extreme cases, the energy generated by the disk is comparable to that generated by the star itself. We can also obtain estimates for the minimum outer radii of these disks, approximately 100 astronomical units (AU), which makes them comparable in size to our solar system (1 AU is the distance between Earth and the Sun). Estimates for the disk masses (M_D) can be obtained by measuring the radiation spectrum at long wavelengths where the amount of radiation observed is directly proportional to M_D. Currently available estimates lie in the range $M_D = 0.01$ to $1.0\ M_\odot$ (Beckwith et al., 1990; Adams, Emerson, and Fuller, 1990). Star/disk systems live in the active phase for approximately 1 million years. The estimated disk properties (radial sizes, masses, and angular momenta) are in good agreement with the disk properties predicted by the protostellar theory. The disk properties are thus perfectly consistent with those required to form a solar system such as ours.

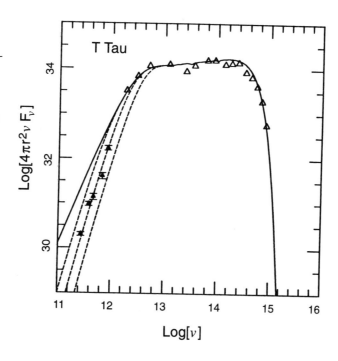

FIGURE 3.9

Observed and theoretical spectral energy distributions of the T Tauri star/disk system. Dashed curves show models with finite disk masses of 0.01, 0.1, and 1.0 $M_\odot$. (Adapted from Adams, Emerson, and Fuller, 1990.)

Summary of the Star-Formation Paradigm

Observations of protostellar candidates are in good agreement with theoretical expectations. The spectral energy distributions calculated from the protostellar theory agree with observed spectra for objects in both the pure infall and the bipolar outflow phases of evolution; hence, infall is probably still taking place in the bipolar phase. The emission from protostellar candidates is observed to be extended in a manner roughly consistent with theoretical expectations. In addition, the spectral energy distributions of T Tauri stars with infrared excesses can be understood as young stars surrounded by circumstellar disks. Some systems have passive disks that reprocess stellar radiation but have no intrinsic luminosity; other systems require appreciable intrinsic disk luminosity. For some sources, estimates for the disk masses ($M_D \sim 0.01$ to $1.0\ M_\odot$) and the disk radii ($R_D \sim 100$ AU) can be obtained.

The current theory of star formation is not yet complete. In particular, we still must understand more about the disk accretion mechanism. Current theory works well for the formation of single stars of low mass (i.e., stars like our Sun), although most stars are members of binary systems (Abt, 1983). The extension of the theory to include the formation of binary systems and stars of higher mass is currently being studied.

PHYSICS OF CIRCUMSTELLAR DISKS

During the protostellar phase of evolution, disk accretion must occur because most infalling material falls onto the disk rather than directly onto the star. In the absence of some mechanism to transfer mass from the disk to the star, the masses of the forming stars would be unrealistically small, the luminosity would be much smaller than that of observed sources, and the disk would become gravitationally unstable. During the T Tauri phase of evolution, disk accretion must also occur to produce the observed intrinsic disk luminosities and to account for the fact that disks seem to disappear on relatively short time scales (Zuckerman, Forveille, and Kastner, 1995). However, the disk accretion mechanism is not fully understood. Two important possible mechanisms are gravitational instabilities and viscous accretion.

A given star/disk system always tends to evolve toward a configuration of lower total energy; this trend holds true for any kind of dissipative mechanism. On the other hand, the system must also conserve its total angular momentum. The lowest energy state accessible to the system is to have essentially all the mass in the central star (so that the material is as far down the gravitational potential well as possible), with one point particle in a large radius orbit to carry all the angular momentum. We expect any star/disk system to evolve toward this minimum energy configuration. This state of affairs is very nearly met within our solar system — almost all the mass resides in the Sun itself, and almost all the angular momentum is carried by the orbital motion of the giant planets (at large radii).

Gravitational Instabilities

Gravitational instabilities in circumstellar disks (Adams, Ruden, and Shu, 1989; Shu, Tremaine, Adams, and Ruden, 1990) might be one route to the formation of giant planets or binary stellar companions of the central star. If the disks are unstable, the net effect is to transfer angular momentum within the disk and possibly lead to disk accretion. We begin with an unperturbed system consisting of a star with an accompanying gaseous disk. We then determine the channels of instability accessible to the system. All disk systems tend to produce spiral patterns; well-known examples of this type of behavior are spiral galaxies and the spiral patterns observed in the rings of Saturn. The growth and behavior of spiral instabilities is mainly determined by three elements: gravity, pressure, and differential rotation.

Recent work has focused on spiral patterns with one spiral arm (in contrast to galaxies, in which spiral patterns generally have two arms). For a typical circumstellar disk, the gravitational potential well is dominated by the star so that the disk rotation curve is nearly **Keplerian**; in other words, the parcels of gas in the disk travel on simple orbits that are analogous to the orbits of planets around the Sun in our solar system. In this case, the particle orbits are simple closed ellipses. In the absence of other forces, the orbits will not change with time. Thus, if the orbits are lined up to form a one-armed pattern, they will stay in that configuration (this is called a purely *kinematic* mode). If the potential is not exactly Keplerian, a relatively small amount of self-gravity in the disk should be sufficient to sustain a one-armed pattern in a circumstellar disk. Hence, spiral patterns with one arm should arise naturally in these systems.

Modes with a single spiral arm are especially interesting in the context of forming star/disk systems because the center of mass of the perturbation does not lie at the system's geometrical center (the star). Formally, this effect gives rise to a new forcing mechanism (a new means of driving wave motions through the disk), which plays an important role in the amplification of these modes. Physically, this effect causes the star to move from the center of mass of the star/disk system; that is, the disk transfers angular momentum to the stellar orbit. The new forcing mechanism is essential for the growth and maintenance of spiral modes with one spiral arm. In fact, this mechanism can be the dominant amplification mechanism for these modes.

From the combined results of many recent studies, we now have a basic understanding of the physics of self-gravitating instabilities in disks. In particular, we can determine the dependence of the growth rates on the underlying physical parameters of the star/disk system. Disks become more stable as the temperature is raised (because of the increase in pressure support). On the other hand, disks tend to become more unstable as the total disk mass is increased (because of the increase in gravitational force).

The growth rates of the instabilities are largest when the disk and star have equal masses ($M_D = M_*$) and decrease rapidly with decreasing relative disk mass. In the optimal case, $M_D = M_*$, the instabilities grow on a time scale comparable to the orbital time at the outer disk edge; that is, the modes can grow on nearly a dynamical time scale. This time scale is typically a few thousand years, which is much shorter than the evolutionary time scale (millions of years) of the disks. Thus, these gravitational instabilities can grow fast enough to significantly affect disk evolution.

For the particular case of spiral modes with one arm, a finite mass threshold exists for strong amplification of the instabilities. The finite threshold implies a critical value of the relative disk mass, the maximum value of the ratio $M_D/(M_* + M_D)$ that is stable to these disturbances. For the simplest case of a perfectly Keplerian disk, the critical ratio has the value

$$\frac{M_D}{(M_* + M_D)} = \frac{3}{4\pi} \qquad (2)$$

Thus, when the disk mass is greater than about one-third the stellar mass, gravitational instabilities grow strongly and the disk is highly unstable. When the disk mass becomes less than about one-third the stellar mass, instabilities can still grow, but at a much slower rate.

Nonlinear Simulations of Star/Disk Systems

To study the nonlinear behavior of gravitational instabilities, one must perform numerical simulations. Results have been obtained from several different hydrodynamic simulations of gravitational instabilities in nearly Keplerian disks (Adams and Benz, 1992; Laughlin and Bodenheimer, 1994; Woodward, Tohline, and Hashisu, 1994). Some numerical simulations of star/disk systems have been performed using a smooth particle hydrodynamics (SPH) computer code. (This calculation is similar to the numerical studies discussed in Chapter 2.) In this type of simulation, computer-generated "particles" are used to represent parcels of gas. All the simulations use an **isothermal equation of state**; from a physical point of view, the use of an isothermal equation of state implicitly assumes that the disk is able to efficiently radiate all energy dissipated during the evolution. Results show that gravitational instabilities can grow strongly in these systems and can produce well-defined spiral patterns (see Figure 3.10). In particular, the growth rates are comparable to the orbital time scale of the outer disk edge and are thus in agreement with the stability calculations described earlier. For disks that are not too far from the condition of stability, spiral instabilities with different numbers of spiral arms ($m = 1,2,3,4$ and higher) develop. As stability is increased (i.e., as the temperature is raised or the mass ratio M_D/M_* is decreased), the relative strength of the

FIGURE 3.10

Result of an SPH simulation of a star/disk system. The main spiral arm has collapsed to form a gravitationally bound clump of gas. In this simulation, the star and disk have equal masses. The time scale for clump formation to occur is a few orbit times of the outer disk edge; this time corresponds to several thousand years for a star/disk system the size of our solar system. (Adapted from Adams and Benz, 1992.)

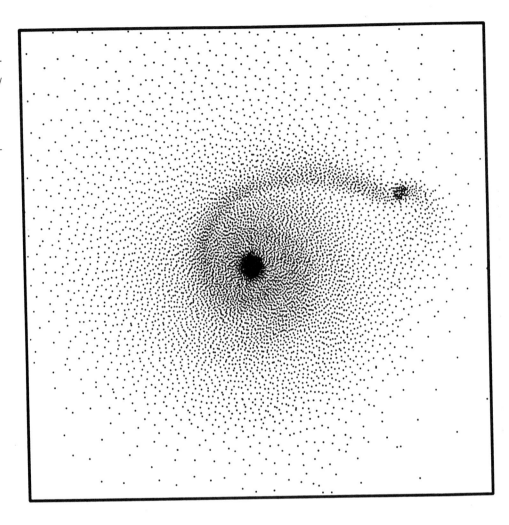

one-armed disturbance increases, although the growth rates of all modes decrease as expected.

A spiral arm can collapse to form a clump of gravitationally bound gas when the instability in the disk becomes sufficiently strong (Figure 3.10). The collapsed clumps typically have masses of about 0.01 M_D and travel on elliptical orbits. The possibility that the clumps survive to form either giant planets or binary companions is especially interesting. If the clumps eventually form binary companions, they must gain mass from the disk faster than does the original star. On the other hand, if the clumps eventually form giant planets, then some mechanism must partially separate the gas from the heavy elements to provide the observed enrichment of the heavy elements. More work is necessary to understand the long-term fate of such clumps.

However, these clumps are not always produced from all types of self-gravitating instabilities (Laughlin and Bodenheimer, 1994; Woodward et al., 1994). As a general trend, to form a gravitationally bound clump in a circumstellar disk, the underlying perturbation must have a density contrast ($\Delta\rho/\rho$) of about 3 or 4. For perturbations with smaller density contrasts, the clumps do not form. Instead, the effects of angular momentum transfer dominate and the spiral modes drive an accretion flow through the disk.

Viscous Evolution of Circumstellar Disks

In the later stages of disk evolution, the disk mass must eventually become small enough that gravitational instabilities turn off, or at least become substantially less important. For the case of intermediate mass disks (i.e., when the disk mass is too small for gravitational effects to dominate, but large enough for planet formation and other disk processes to take place), we expect circumstellar disks to evolve through the action of viscosity. Such systems are generally known as *viscous accretion disks*. Any source of fluid viscosity (which is essentially a frictional force) creates both energy dissipation and transfer of angular momentum; these are precisely the two physical effects that must occur for disk accretion to take place.

The most important issue in this picture is the source of the fluid viscosity. Ordinary molecular viscosity is always present and arises from frictional forces between molecules; however, this viscosity is much too small to be of astrophysical importance. Thus, a source of anomalous viscosity must somehow be generated. Many different mechanisms have been proposed to generate turbulence in circumstellar disks; turbulence, in turn, leads to small-scale motions that allow for the dissipation of energy and, hence, an effective viscosity. One of the leading ideas for generating turbulence is from convection in the vertical direction in the disk (Lin and Papaloizou, 1985; Ruden and Lin, 1986). However, at the present time, the source of anomalous viscosity remains controversial.

The evolution of viscous accretion disks is governed by the laws of fluid dynamics. In the limit that viscous forces drive the evolution of the disk, the behavior is described by a time-dependent diffusion equation (Lüst, 1952; Lynden-Bell and Pringle, 1974; Lin and Papaloizou, 1985). In other words, disk evolution occurs through a viscous diffusion process. As in any diffusion process, the net result is for the system (here, the disk) to spread out. In this case, the inner parts of the disk move farther inward and some disk material is accreted onto the star in the center. On the other hand, the outer parts of the disk gain angular momentum and spread farther outward. As this process continues, the disk mass becomes smaller relative to the star, the surface density of the disk decreases, and the outer radius of the disk increases. Thus, this process makes a star/disk system evolve toward the current state of our own solar system, with most of the mass in the central star and most of the angular momentum carried by objects at large radii (the planets). The evolutionary time scale depends on the size of the viscosity.

Additional Disk Processes

In addition to the mechanisms for angular momentum transport outlined above, other possibilities exist. For example, if the magnetic field strength is large enough, magnetic torques can lead to the transport of angular momentum (Hoyle, 1960; Hayashi, 1981; Stepinsky and Levy, 1990). One way to view magnetic effects is to define an effective magnetic viscosity proportional to B^2, where B is the magnetic field strength. The magnetic viscosity drives disk evolution in a manner roughly analogous to the viscous evolution scenario described above. In this case, however, the strength of the magnetic field determines the size of the viscosity and, hence, the time scale for evolution.

On the other hand, if the magnetic field is relatively weak, a new type of instability can arise (Balbus and Hawley, 1991); very roughly, the magnetic field provides a means of transferring energy from the orbital motion in the disk into small-scale motions. This process can occur whenever the energy density of the magnetic field is less than the thermal energy density. The resulting instability leads to turbulence in the disk and may also provide a mechanism for angular momentum transport.

Tidal effects can also be important in circumstellar disks; these effects can arise when secondary bodies (either planets or binary companions) exert gravitational forces on the moving parcels of gas in the disk. If planets form in the disk, they can clear out gaps in much the same way as moons of Saturn clear gaps in its rings (Lin and Papaloizou, 1986a,b). Because most stars are found in binary systems, binary companions should be present in many star/disk systems. They can also transport angular momentum through the disk by tidal torques.

Summary of Disk Processes

Many different processes can lead to angular momentum transport and hence accretion in circumstellar disks. Self-gravitating instabilities can grow on nearly a dynamical time scale (typically ~ 10^3 yr, the orbital time scale at the outer disk edge), which is much shorter than the evolutionary time scale of these systems (10^5 to 10^6 yr). Computer simulations show that these instabilities can grow well into the nonlinear regime; in other words, the density of the growing spiral arms can become larger than the initial density in the disk. For perturbations that become sufficiently large, small clumps of gravitationally bound gas can form out of the disk. For less successful perturbations that halt their growth at lower amplitudes, the instabilities can drive angular momentum transport and disk accretion. Thus, gravitational instabilities may play a role in the disk accretion processes, the formation of binary companions from the disk, or both.

In addition to gravitational instabilities, other physical processes can lead to disk accretion. We expect that self-gravitating instabilities are most important early in the evolution of the disk (when the disk mass is relatively large). At sufficiently late times, the disk mass must eventually become small enough that gravitational instabilities shut off and some type of viscous accretion process takes over. The presence of any type of viscosity leads to viscous diffusion of the disk. In other words, the disk spreads out and produces an accretion flow from the disk onto the star. At sufficiently long times, the star/disk system is driven to a final state much like that of our solar system — most of the mass ends up in the central star and most of the angular momentum resides in the orbital motion of material (perhaps planets) at large radii.

PLANET FORMATION

The hypothesis that planets form within the circumstellar disks that surround young stellar objects — *the nebular hypothesis* — was first stated two centuries ago (Kant, 1755; Laplace, 1796) and continues to be the most viable scenario for planet formation. The disk properties provide the initial conditions for the process of planet formation.

Very roughly, we can identify two conceptually different ways for planets to form within a circumstellar disk. The first of these processes, accumulation of **planetesimals**, assumes that planets form by the gradual accretion of small, rock-like bodies. In this case, the planets form "from the bottom up." Alternatively, planets could also form through a gravitational instability in the disk. In this case, the circumstellar disk becomes gravitationally unstable and breaks into secondary bodies that become the planets. In the second situation, the planets form "from the top down." Although both planet formation scenarios have some difficulties, the accretion of planetesimals is generally thought to be more likely.

The problem of planet formation is on a somewhat different footing than the overall problem of star formation. As discussed above, we now have many examples of observed objects which we think are forming or newly formed stars. However, we have not yet directly observed planets in other solar systems; we have direct information only about the planets associated with our Sun.

Our solar system contains two different types of planets. The inner four, Mercury, Venus, Earth, and Mars, are known as the *terrestrial planets*. They are composed primarily of heavy elements (i.e., elements heavier than hydrogen and helium, which make up 99% of the mass of the Sun). The next four planets, Jupiter, Saturn, Uranus, and Neptune, are known as the *giant planets* or the *Jovian planets*. They are much more massive than the terrestrial planets and contain substantial amounts of hydrogen and helium. They do not, however, contain as much hydrogen and helium as the Sun; they are enriched in heavy elements, some very much so. The giant planets also differ from stars (and the Sun) in that they contain solid rocky cores in their centers. The elemental abundances of the different planets place important constraints on how they must have formed. In addition, our solar system contains many other bodies, including asteroids, the planet Pluto, and comets.

The circumstellar disk from which the planets formed must have contained a substantial amount of mass. Clearly, the mass of this disk must have been at least as large as the total mass of all our planets and other solar system bodies. However, because these bodies contain relatively more heavy elements than the Sun itself, the initial total disk mass must have been at least as large as the total mass of all solar system bodies when augmented to solar abundances, that is, when the missing hydrogen and helium are added back into the total. The augmented disk would have had a mass of about 0.01 $M_\odot$, known as the *minimum mass solar nebula*.

Formation of Planets by Accumulation

Let us first consider planet formation through the accumulation of small solid bodies. In disks that form planets, most of the heavy elements are found initially in the form of dust grains. In the interstellar medium (before gas and dust become incorporated into star and disk), dust grains are very small, with typical radii about 10^{-5} cm. In the denser environment of a circumstellar disk, the grains are expected to become somewhat larger, but remain mostly microscopic in size (small, compared with 1 cm). Thus, the process of planet formation begins with tiny dust grains and somehow produces huge objects with characteristic sizes of about 10^9 cm.

Dust grains accumulate on a fairly short time scale into large rock-like bodies called *planetesimals*. In the absence of turbulence, the dust grains settle to the midplane of the disk because they do not feel the same pressure force as the gas. In this case, the resulting thin layer of dust becomes gravitationally unstable and breaks up into planetesimals. The planetesimals are basically large rocks with sizes about 1 km, that is, about the size of an asteroid. In other words, the planetesimals are much larger than the original dust grains, but are still much smaller than a planet. The picture of planetesimal production is complicated by the presence of turbulence, which prevents the dust grains from settling all the way to disk midplane and inhibits gravitational instability. In this case, sticking together of dust grains becomes important. However, the net result is similar in that large rocks of roughly asteroid size and mass can be produced quickly. The time scale for the production of planetesimals is about 10^4 years, which is short compared with the expected lifetime of the disk (roughly 10^7 yr).

The next stage of planet formation involves the accumulation of planetesimals into the planets themselves. The terrestrial planets are composed mostly of the rocky material that makes up the planetesimals, so their formation can be understood entirely in terms of planetesimal accumulation. However, the giant planets also contain substantial amounts of the lighter gases. For the latter type of composition to arise, the planetesimals must accumulate into a massive rocky core that subsequently accretes gas from the circumstellar disk. Giant planets thus require an additional stage of formation.

The accumulation of planetesimals into planets takes considerably longer than the initial buildup of the planetesimals. One relevant time scale for this process is the orbit time at the radial position of the forming planet. This time scale is 1 year at the radius of Earth's orbit, 1/4 year at the radius of Mercury's orbit, and 164 years at the radius of Neptune's orbit. Thus, the natural clock runs considerably slower in the outer solar system than it does in the inner parts. Another important effect is that the density of the disk (and, hence, the number density of the planetesimals) decreases with radius. This effect makes the likelihood of planetesimal collisions smaller in the outer solar system. As a result, it takes a long time for planetesimals to accumulate in the outer solar system.

The time scales to build up planet-sized bodies from planetesimals remain poorly understood. As a rough estimate, under the most favorable circumstances, the cores of the giant planets can be produced in about 10^6 years. Because the gas in the disk has a lifetime at least this long, the cores can accrete enough gaseous material to account for the observed compositions of the giant planets. Thus, this scenario for planet formation basically works. Other calculations, however, suggest that the time scale for building the cores of the outer planets could be as long as 10^8 years, longer than the expected lifetime of the gaseous disk. These issues are still being studied.

Formation of Planets by Gravitational Instability

Planets may also form through a gravitational instability in the disk. Gravitational condensation alone cannot account for the composition of the planets. Gravity does not distinguish between gas and rocky particles. Thus, any objects formed through the action of gravity alone are expected to have the same composition as the Sun. However, it is possible for the outer giant planets to have formed from a gravitational condensation in the disk, provided that some other mechanism can produce the observed enrichment in heavy elements.

As discussed already, circumstellar disks can be unstable if their temperatures are low enough and their masses are large enough. Furthermore, many observed star/disk systems seem to live near this point of gravitational instability. The natural mass scale for a gravitationally bound object that forms in a disk can be shown to be roughly 0.01 $M_\odot$ for the case of a star and disk with nearly equal masses (for solar-type stars); this mass is more than enough to form a giant planet. The relevant time scale for gravitational condensation is comparable to the orbital time scales of 10 to 10^3 years, much shorter than the accumulation time scale. Thus, the gravitational instability mechanism has both a time scale and a mass scale that are appropriate for the formation of giant planets.

The difficulty in this scenario of planet formation is in understanding how the heavy elements become enriched and how most heavy elements end up in the cores of the planets. Dust grains, which contain the heavy elements, naturally tend to settle toward the center of a self-gravitating protoplanet because the grains feel less force from pressure than the surrounding gas. However, the time scale for the dust-settling process is generally thought to be too long (more than 10^8 yr) to account for the observed composition and structure of the giant planets. Other mechanisms to provide dust separation have also been proposed (e.g., dust settling in vortices), but this issue also remains under study.

THE INITIAL MASS FUNCTION

The IMF is perhaps the most fundamental output of the star formation process. The IMF is the distribution of masses of a given stellar population at the moment of birth. A detailed knowledge of the IMF is required to understand galaxy formation, the chemical evolution of galaxies, the evolution of the interstellar medium, and other important astronomical issues. Unfortunately, at the present time, we remain unable to calculate the initial mass function from a priori considerations (Zinnecker, McCaughrean, and Wilking, 1992).

The IMF Observed

We can, however, empirically determine the IMF today in our galaxy. As a first approximation (Salpeter, 1955), the number of stars born with masses in the range M_* to $M_* + dM_*$ is given by the simple power-law relation

$$f(M_*)dM_* \sim M_*^{-\beta} dM_* \tag{3}$$

where the index $\beta = 2.35$ for stars in the mass range $M_* = 0.4$ to $10\ M_\odot$. Thus, many more stars are born with small masses than with larger masses. To make this point precise, stars with masses 10 times that of our Sun are about 220 times less numerous than stars with the mass of the Sun. More recent work (Miller and Scalo, 1979; Scalo, 1986) suggests that the mass distribution is somewhat more complicated than a simple power-law and may have two local maxima: a primary maximum at M_* about $0.3\ M_\odot$ and a (weaker) secondary maximum at M_* about $1.2\ M_\odot$. In addition, the distribution seems to flatten out (and perhaps turn over) at the lowest masses ($M_* \sim 0.1\ M_\odot$).

Stars can only exist in a finite range of masses. Stellar objects with masses less than about $0.08\ M_\odot$ cannot produce central temperatures hot enough for the fusion of hydrogen to take place. Objects with masses less than this hydrogen-burning limit are known as *brown dwarfs*. On the other end of the possible mass range, stars with masses greater than about $100\ M_\odot$ cannot exist because they are unstable. Thus, stars are confined to the rather narrow mass range

$$0.08\ M_\odot \leq M_* \leq 100\ M_\odot \tag{4}$$

This mass range is much smaller than the conceivable mass range. Stars form in galaxies that have masses of about 10^{11} $M_\odot$ and stars are made up of hydrogen atoms that can have masses of about 10^{-24} g approximating 10^{-57} $M_\odot$. Thus, galaxies could build objects anywhere in the mass range from 10^{-57} $M_\odot$ to 10^{11} $M_\odot$, a factor of 10^{68} in mass scale! Yet stars live in the mass range given by Equation 4, which allows stellar masses to differ from each other by a factor of only 10^3.

The IMF as a Two-Phase Process

In considering the determination of the IMF, we can conceptually divide the process into two basic subproblems: the spectrum of initial conditions produced by the interstellar medium, and the transformation between a given set of initial conditions and the properties of the final (formed) star.

The Spectrum of Initial Conditions To understand the spectrum of initial conditions, we must understand all the relevant processes that occur in the formative medium (molecular clouds in the case of present-day star formation). In particular, we must understand the mechanism that selects the distribution of properties for molecular cloud cores (these cores provide the initial conditions for protostellar collapse). More work on the formation of substructure within molecular clouds is needed before we can begin to predict the distribution of core properties from a priori considerations.

Although we cannot predict the spectrum of initial conditions, we can measure the conditions in molecular clouds. For example, molecular clouds seem to break up into smaller units, which we denote as clumps. Many groups (Blitz, 1993) have studied the distribution of clump masses in molecular clouds and have found power-law forms; that is, the distribution of the number of clumps with mass M_{cl} obeys the relation

$$\frac{dN_{cl}}{dM_{cl}} \sim M_{cl}^{-p} \qquad (5)$$

The index p is approximately 3/2 for almost all molecular clouds. Molecular clouds apparently tend to fragment into pieces (clumps) in such a way that the mass distribution of the clumps is a simple power-law. Furthermore, all molecular clouds seem to fragment with nearly the same power-law. Thus, some type of universal fragmentation process seems to be at work. Molecular clouds are highly nonuniform; clumpiness and structure exist on many different size scales (Figure 3.3). This result is important for star formation because no single characteristic density exists for these clouds.

Semi-Empirical Mass Formula To understand the transformation between initial conditions and the final masses of the stars produced, we must know how the stellar outflow stops the inflow and separates the star/disk system from its molecular environment. This process is still being studied. The basic idea is that the final mass of a star is determined by the point at which the momentum of the outflow becomes greater than the momentum of the infalling material. This crossover point always occurs. The total flux of momentum of the infall is generally a constant in time, but the infall that lands directly on the star itself decreases with time. On the other hand, the strength of the outflow is correlated with the total energy output of the system and hence grows with time (the mass of the system grows with time and more massive objects have greater energy output). As a result, if one waits long enough, the outflow strength will always grow larger than that of the infall.

The above argument leads to the scaling law

$$L_* M_*^2 = \chi \frac{a^{11}}{G^3 \Omega^2} \tag{6}$$

where the parameter χ is a pure number about 250 (Shu, Lizano, and Adams, 1987). This expression provides a relation between the final properties of the newly formed star and quantities that describe the initial conditions for star formation. The left-hand side of the equation contains the final system properties, the luminosity L_* and the mass M_*, evaluated at the point of evolution when the wind is capable of reversing the direct infall. The right-hand side of the equation contains the initial conditions: the sound speed a (which is determined by the initial core temperature) and the rotation rate Ω of the initial molecular cloud core. If we use "typical" values for present-day clouds (e.g., $a = 0.2$ km s^{-1} and $\Omega \sim 3 \times 10^{-14}$ rad s^{-1}), we obtain stellar masses of approximately 0.5 $M_\odot$, which is the typical mass of stars forming in regions with these properties. Thus, in spite of its highly idealized nature, the scaling relation provides reasonable estimates for the masses M_* of forming stars as a function of initial conditions (a, Ω). More importantly, the scaling relation shows that the masses of forming stars should increase with sound speed (or, equivalently, with temperature) and should decrease with rotation rate Ω.

In the interpretation of the above transformation, several points must be kept in mind. The right-hand side of the equation depends rather sensitively on the sound speed ($\sim a^{11}$) and hence the temperature ($\sim T^{11/2}$). However, to evaluate the left-hand side of the equation, we must use a luminosity-mass relation for stars; for low-mass stars on the main sequence, L_* is proportional to M_*^4, whereas for high-mass stars L_* is proportional to M_*. We thus obtain (very crude) scaling relations of the form M_* proportional to $T^{11/12}$ for low-mass stars and M_* proportional to $T^{11/3}$ for high-mass stars. Much uncertainty has been encapsulated in the parameter χ, which should really be considered as a complicated function of all the environmental parameters. We have also characterized the initial conditions by only two physical variables (a and Ω), but much more complicated initial states are possible.

A Scaling Relation for the IMF

We can now construct an IMF. First, the presence of turbulence in molecular clouds leads to a scaling relation between the small scale velocities Δv and the mass of the clump, that is, M_{cl} is proportional to $(\Delta v)^4$. In other words, larger clumps tend to have more energetic turbulent motions and hence larger values of Δv. Next, we interpret the velocity Δv as the effective transport speed a, which determines the initial condition for star formation. If we now combine this result with the clump mass spectrum of Equation 5 and the semiempirical mass formula of Equation 6, we obtain an IMF of the form

$$\frac{dN}{dM_*} = \frac{dN}{dM_{cl}} \frac{dM_{cl}}{dM_*} \sim M_*^{-\beta} \tag{7}$$

Thus, this simple argument produces a power-law IMF with an index in the range $\beta = 1.6$ to 2.1. This result compares reasonably well with the observed power-law index of the IMF which has β about 2.35 (Salpeter, 1955; Silk, 1995).

The basic logic of this argument can be summarized as follows. Molecular clouds produce a distribution of initial conditions for star formation. In the simplest picture considered here, the clouds produce a distribution of clump masses. Because larger clumps have larger effective sound speeds caused by turbulence and other small-scale physical processes, the distribution of clump masses can be converted into a distribution of effective sound speeds, which represents the initial conditions for star formation. We then use the idea that outflows help determine the final masses of forming stars to find a transformation between the initial conditions and the final stellar masses. Finally, the distribution of initial conditions can be transformed into the distribution of stellar masses, and we thereby obtain the IMF.

SUMMARY AND DISCUSSION

This chapter outlines the current theory of star formation in molecular clouds. We can conceptually divide the star formation process into four stages (see Figure 3.2). In the first stage, molecular clouds produce small dense regions called molecular cloud cores. In the next stage, a core collapses to form a star/disk system that is deeply embedded in an infalling envelope of dust and gas. The star/disk system develops a powerful outflow (wind) in the next stage. Our current theory suggests that stars, in part, determine their own masses through the action of the powerful outflows. In other words, the outflows help separate a newly formed star/disk system from its molecular environment. In the fourth, final stage of star formation, the optically revealed star evolves into a configuration capable of sustaining the nuclear fusion of hydrogen. Also during this stage, the circumstellar disk transfers a substantial portion of its mass onto the star, produces planets, or both. The entire star formation process takes place on a time scale of only a few million years.

This theory of star formation has many successes, especially for stars of low mass. In particular, the theory predicts both spectral energy distributions and spatial distributions of emission in good agreement with observations. The presence of circumstellar disks, which often must have substantial activity, can explain the spectral appearance of newly formed stars with infrared excesses. The estimated masses and radii of the circumstellar disks are consistent with those properties necessary to form solar systems similar to our own.

In spite of its successes, the theory of star formation remains incomplete. One important unresolved issue is the mechanism that leads to accretion through the disks. We now understand the basic physics by which gravitational instabilities and viscosity can lead to disk accretion. For the former, however, we must understand the long-term nonlinear evolution. For the latter, we must understand the source of the viscosity. In addition, the theory still must be extended to include the formation of high-mass stars and the formation of binary companions. We have just begun to calculate and predict the initial mass function for forming stars as a function of environmental parameters.

We began this chapter by noting that star formation and stellar evolution can be considered as a war between gravity and entropy. The battle takes place on many different size scales: in molecular clouds, in molecular cloud cores, in circumstellar disks, and within the stars themselves. The force of gravity must win at least a partial victory for stars and planets to form. However, the process of star formation ends with the production of a main sequence star, which represents a truce between the two warring parties — in these stars, the gravitational forces are exactly balanced by pressure forces generated through nuclear processes. The apparent state of peace is only temporary. Stars eventually run out of their nuclear fuel and must readjust their configurations; in other words, the war between gravity and entropy starts up again at the end of a star's life. The final readjustment can lead to violent explosions and the production of exotic stellar objects such as white dwarfs, neutron stars, and black holes (see Chapter 4).

REFERENCES

Abt, H. A. 1983. Normal and abnormal binary frequencies. *Annual Review of Astronomy and Astrophysics 21*: 343–372.

Adams, F. C. 1990. The L–A_V diagram for protostars. *Astrophysical Journal 363*: 578–588.

Adams, F. C., and Benz, W. 1992. Gravitational instabilities in circumstellar disks and the formation of binary companions. In: H. McAlister (Ed.), *Complementary Approaches to Binary and Multiple Star Research* (IAU Coll. No. 135) (Provo, Utah: Astronomical Society of the Pacific), pp. 185–194.

Adams, F. C., Emerson, J. P., and Fuller, G. A. 1990. Submillimeter photometry and disk masses of T Tauri disk systems. *Astrophysical Journal 357*: 606–620.

Adams, F. C., Lada, C. J., and Shu, F. H. 1987. Spectral evolution of young stellar objects. *Astrophysical Journal 312*: 788–806.

Adams, F. C., Lada, C. J., and Shu, F. H. 1988. The disks of T Tauri stars with flat infrared spectra. *Astrophysical Journal 326*: 865–883.

Adams, F. C., Ruden, S. P., and Shu, F. H. 1989. Eccentric gravitational instabilities in nearly Keplerian disks. *Astrophysical Journal 347*: 959–975.

Adams, F. C., and Shu, F. H. 1986. Infrared spectra of rotating protostars. *Astrophysical Journal 308*: 836–853.

Appenzeller, I., and Mundt, R. 1989. T Tauri stars. *Astronomy and Astrophysics Reviews 1*: 291–324.

Arons, J., and Max, C. 1975. Hydromagnetic waves in molecular clouds. *Astrophysical Journal 196*: L77–L82.

Balbus, S. A., and Hawley, J. F. 1991. A powerful local shear instability in weakly magnetized disks. I. Linear analysis. *Astrophysical Journal 376*: 214–222.

Beckwith, S., Sargent, A. I., Chini, R., and Gusten, R. 1990. A survey for circumstellar disks around young stars. *Astronomical Journal 99*: 924–945.

Blitz, L. 1993. Giant molecular clouds. In: E. Levy and M. S. Mathews (Eds.), *Protostars and Planets III* (Tucson: University of Arizona Press), pp. 125–162.

Butner, H. M., Evans, N. J., Lester, D. F., Levreault, R. M., and Strom, S. E. 1991. Testing models of low-mass star formation: High resolution far-infrared observations of L1551 IRS5. *Astrophysical Journal 376*: 636–653.

Calvet, N., Hartmann, L. W., Kenyon, S. J., and Whitney, B. A. 1994. Flat spectrum T Tauri stars: The case for infall. *Astrophysical Journal 434*: 330–340.

Cassen, P., and Moosman, A. 1981. On the formation of protostellar disks. *Icarus 48*: 353–376.

Goodman, A. A. 1990. Interstellar magnetic fields: An observational perspective. Ph.D. thesis, Harvard University.

Hayashi, C. 1981. Structure of the solar nebula, growth and decay of magnetic fields and effects of magnetic and turbulent viscosities on the nebula. *Progress of Theoretical Physics* (Suppl.) *70*: 35–53.

Heiles, C. H., Goodman, A. A., McKee, C. F., and Zweibel, E. G. 1993. Magnetic fields in star-forming regions: Observations. In: E. Levy and M. S. Mathews (Eds.), *Protostars and Planets III* (Tucson: University of Arizona Press), pp. 279–326.

Houlahan, P., and Scalo, J. 1992. Recognition and characterization of hierarchical interstellar structure. II. Structure tree statistics. *Astrophysical Journal 393*: 172–187.

Hoyle, F. 1960. On the origin of the solar system. *Quarterly Journal of the Royal Astronomical Society 1*: 28–55.

Kant, I. 1755. *Allegmeine Naturgeschichte und Theorie des Himmels*. Germany. English translation: Kant, I. 1986. *Universal Natural History and Theory of the Heavens*, trans. by S. K. Jaki (Edinburgh: Scottish Academic Press).

Kenyon, S., and Hartmann, L. 1987. Spectral energy distributions of T Tauri stars: Disk flaring and limits on accretion. *Astrophysical Journal 323*: 714–733.

Kenyon, S. J., Calvet, N., and Hartmann, L. W. 1993. The embedded young stars in the Taurus-Auriga molecular cloud. I. Spectral energy distributions. *Astrophysical Journal 414*: 676–694.

Lada, C. J. 1985. Cold outflows, energetic winds, and enigmatic jets around young stellar objects. *Annual Review of Astronomy and Astrophysics 23*: 267–317.

Lada, C. J., and Shu, F. H. 1990. The formation of sunlike stars. *Science 1111*: 1222–1233.

Ladd, E. F., Adams, F. C., Casey, S., Davidson, J. A., Fuller, G. A., Harper, D. A., Myers, P. C., and Padman, R. 1991. Far infrared and submillimeter wavelength observations of star forming dense cores. II. Spatial distribution of continuum emission. *Astrophysical Journal 382*: 555–569.

Laplace, P. S. 1796. *Exposition du Système du Monde*. (Paris). English translation in: Knickerbocker, W. S. 1927. *Classics of Modern Science* (Boston: Beacon Press).

Laughlin, G. P., and Bodenheimer, P. 1994. Nonaxisymmetric evolution in protostellar disks. *Astrophysical Journal 436*: 335–354.

Lin, D.N.C., and Papaloizou, J.C.B. 1985. On the dynamical origin of the solar system. In: D. C. Black and M. S. Mathews (Eds.), *Protostars and Planets II* (Tucson: University of Arizona Press), pp. 981–1072.

Lin, D.N.C., and Papaloizou, J.C.B. 1986a. On the tidal interaction between protoplanets and the primordial solar nebula. II. Self-consistent nonlinear interaction. *Astrophysical Journal 307*: 395–409.

Lin, D.N.C., and Papaloizou, J.C.B. 1986b. On the tidal interaction between protoplanets and the primordial solar nebula. III. Orbital migration of protoplanets. *Astrophysical Journal 309*: 846–857

Lizano, S., and Shu, F. H. 1989. Molecular cloud cores and bimodal star formation. *Astrophysical Journal 342*: 834–854.

Lüst, R. 1952. Die entwicklung einer um einen zeutralkörper rotierenden gasmasse. I. Loesungen de hydrodynamischen gleichungenmit turulenter reibung. *Zeitschrift für Naturforschung 7a*: 87–98.

Lynden-Bell, D., and Pringle, J. E. 1974. The evolution of viscous disks and the origin of the nebular variables. *Monthly Notices of the Royal Astronomical Society 168*: 603–637.

Miller, G. E., and Scalo, J. M. 1979. The initial mass function and stellar birth rate in the solar neighborhood. *Astrophysical Journal* (Suppl.) *41*: 513–547.

Mouschovias, T. 1976. Nonhomologous contraction and equilibria of self-gravitating magnetic interstellar clouds embedded in an intercloud medium: Star formation. I. Formulation of the problem and method of solution. *Astrophysical Journal 206*: 753–767.

Mouschovias, T. 1978. Formation of stars and planetary systems in magnetic interstellar clouds. In: T. Gehrels (Ed.), *Protostars and Planets* (Tucson: University of Arizona Press), pp. 209–242.

Mouschovias, T., and Spitzer, L. 1976. Note on the collapse of magnetic interstellar clouds. *Astrophysical Journal 210*: 326–327.

Myers, P. C. 1985. Molecular cloud cores. In: D. C. Black and M. S. Mathews (Eds.), *Protostars and Planets II* (Tucson: University of Arizona Press), pp. 81–103.

Myers, P. C., Fuller, G. A., Mathieu, R. D., Beichman, C. A., Benson, P. J., Schild, R. E., and Emerson, J. P. 1987. Near-infrared and optical observations of IRAS sources in and near dense cores. *Astrophysical Journal 319*: 340–357.

References

Myers, P. C., and Fuller, G. A. 1992. Density structure and star formation in dense cores with thermal and nonthermal motions. *Astrophysical Journal 396*: 631–648.

Myers, P. C., and Goodman, A. A. 1988. Magnetic molecular clouds: Indirect evidence for magnetic support and ambipolar diffusion. *Astrophysical Journal 329*: 392–405.

Nakano, T. 1984. Contraction of magnetic interstellar clouds. *Fundamentals of Cosmic Physics 9*: 139–232.

Palla, F., and Stahler, S. W. 1990. The birthline for intermediate mass stars. *Astrophysical Journal 360*: L47–L50.

Ruden, S. P., and Lin, D.N.C. 1986. The global evolution of the primordial solar nebula. *Astrophysical Journal 308*: 883–901.

Rydgren, A. E., and Zak, D. S. 1987. On the spectral form of the infrared excess component in T Tauri systems. *Publications of the Astronomical Society of the Pacific 99*: 141–145.

Rucinski, S. M. 1985. IRAS observations of T Tauri and post-T Tauri stars. *Astronomical Journal 90*: 2321–2330.

Salpeter, E. E. 1955. The luminosity function and stellar evolution. *Astrophysical Journal 121*: 161–167.

Scalo, J. M. 1986. The stellar initial mass function. *Fundamentals of Cosmic Physics 11*: 1–278.

Scalo, J. M. 1987. Theoretical approaches to interstellar turbulence. In: D. J. Hollenbach and H. A. Thronson (Eds.), *Interstellar Processes* (Dordrecht: Reidel), pp. 349–392.

Shu, F. H. 1977. Self-similar collapse of isothermal spheres and star formation. *Astrophysical Journal 214*: 488–497.

Shu, F. H. 1983. Ambipolar diffusion in self-gravitating isothermal layers. *Astrophysical Journal 273*: 202–213.

Shu, F. H. 1985. Star formation in molecular clouds. In: H. van Woerden, W. B. Burton, and R. J. Allen (Eds.), *The Milky Way* (IAU Symposium No. 106) (Dordrecht: Reidel), pp. 561–565.

Shu, F. H., Adams, F. C., and Lizano, S. 1987. Star formation in molecular clouds: Observation and theory. *Annual Review of Astronomy and Astrophysics 25*: 23–81.

Shu, F. H., Lizano, S., and Adams, F. C. 1987. Star formation in molecular cloud cores. In: M. Peimbert and J. Jugaku (Eds.), *Star Forming Regions* (IAU Symposium No. 115) (Dordrecht: Reidel), pp. 417–434.

Shu, F. H., Lizano, S., Ruden, S. P., and Najita, J. 1988. Mass loss from rapidly rotating magnetic protostars. *Astrophysical Journal 328*: L19–L23.

Shu, F. H., Najita, J., Wilkin, F., Ruden, S. P., and Lizano, S. 1994. Magnetocentrifugally driven flows from young stars and disks. I. A generalized model. *Astrophysical Journal 429*: 781–796.

Shu, F. H., Tremaine, S., Adams, F. C., and Ruden, S. P. 1990. SLING amplification and eccentric gravitational instabilities in gaseous disks. *Astrophysical Journal 358*: 495–514.

Silk, J. 1995. A theory for the initial mass function. *Astrophysical Journal 438*: L41–L44.

Stahler, S. W. 1983. The birthline of low-mass stars. *Astrophysical Journal 274*: 822–829.

Stahler, S. W. 1988. Deuterium and the stellar birthline. *Astrophysical Journal 332*: 804–825.

Stahler, S. W., Shu, F. H., and Taam, R. E. 1980. The evolution of protostars. I. Global formulation and results. *Astrophysical Journal 241*: 637–654.

Stepinsky, T. F., and Levy, E. H. 1990. Dynamo magnetic field induced angular momentum transport in protostellar nebulae: The "minimum mass protosolar nebula." *Astrophysical Journal 350*: 819–826.

Terebey, S., Shu, F. H., and Cassen, P. 1984. The collapse of the cores of slowly rotating isothermal clouds. *Astrophysical Journal 286*: 529–551.

Walker, C. K., Adams, F. C., and Lada, C. J. 1990. 1.3 millimeter continuum observations of cold molecular cloud cores. *Astrophysical Journal 349*: 515–528.

Wiseman, J. J., and Adams, F. C. 1994. A quantitative analysis of IRAS maps of molecular clouds. *Astrophysical Journal 435*: 708–721.

Wood, D.O.S., Myers, P. C., and Daugherty, D. A. 1994. IRAS images of nearby dark clouds. *Astrophysical Journal* (Suppl.) *95*: 457–501.

Woodward, J. W., Tohline, J. E., and Hashisu, I. 1994. The stability of thick self-gravitating disks in protostellar systems. *Astrophysical Journal 420*: 247–267.

Zinnecker, H., McCaughrean, M. J., and Wilking, B. A. 1993. The initial stellar population. In: E. Levy and M. S. Mathews (Eds.), *Protostars and Planets III* (Tucson: University of Arizona Press), pp. 429–496.

Zuckerman, B., and Palmer, P. 1974. Radio emission from interstellar molecules. *Annual Review of Astronomy and Astrophysics 12*: 279–313.

Zuckerman, B., Forveille, T., and Kastner, J. H. 1995. Inhibition of giant-planet formation by rapid gas depletion around young stars. *Nature 373*: 494–496.

CHAPTER 4

STELLAR EXPLOSIONS, NEUTRON STARS, AND BLACK HOLES

Alexei V. Filippenko

INTRODUCTION

Our solar system formed about 4.6 billion years ago from a **nebula**, a giant cloud of gas and dust in the cosmos (see Chapter 3 for details of star formation). Since its birth, our Sun, a typical star, has been generating energy through the **nuclear fusion** of hydrogen to helium (see Chapter 5). Two neutrons and two protons tightly bound in a helium nucleus have slightly less mass than the four original protons (hydrogen nuclei), and this mass deficit (m) is converted to energy (E) according to Albert Einstein's famous equation, $\boldsymbol{E = mc^2}$, where c is the speed of light in a vacuum. In our Sun's case, about 700 million tons of hydrogen are fused each second, at a temperature of about 15 million **kelvins** (K), but the fuel supply is vast; the Sun will continue this process for another 5 billion years, increasing in brightness only slightly as the composition of its interior gradually changes. This **main-sequence** phase of a star lasts as long as there is hydrogen in its **core**.

Eventually, however, the core (~ 10% of the total mass) consists almost entirely of helium, which requires much higher temperatures for fusion into heavier elements. As the inert core loses heat, it contracts because of the pull of gravity; consequently, energy is liberated (just as a dropped ball picks up speed). Roughly half the energy escapes from the star, but the other half heats the core and the surrounding layer of hydrogen; the inner part of the hydrogen shell therefore continues to fuse into helium, but at an accelerated rate, and thereby increases the mass of the core. The excess radiation produced by the core contraction and accelerated fusion causes the envelope of the

---

This chapter is dedicated by the author to the memory of Willy Fowler of the California Institute of Technology.

star to expand by an enormous factor, and the star becomes a luminous **red giant** with a relatively cool surface. When our Sun eventually goes through this phase in 5 to 6 billion years, its diameter will swell to nearly half the orbit of Mercury. The Sun will become so bright that the Earth's surface will literally be fried, and all life will certainly be destroyed.

If the star's initial mass is at least half the Sun's mass, that is, 0.5 **solar masses** (0.5 $M_\odot$), the temperature of the core grows until it reaches about 100 million K, at which point helium nuclei begin to fuse into carbon and oxygen (see Chapter 5). This phase is relatively short-lived, however, because the amount of energy liberated per **nuclear reaction** is much less than that from hydrogen fusion. After roughly 1 billion years, the core consists of carbon and oxygen nuclei, which are not able to fuse to heavier elements because the temperature is too low. As was previously the case with the helium core, the carbon-oxygen core gravitationally contracts, becoming hotter and releasing energy. Helium and hydrogen fusion proceed more vigorously in shells surrounding the core, and the star swells up to form an even larger red giant.

At this point, the star's outer layers become unstable; they can be expelled rather gently through **stellar winds** and a series of "cosmic burps" to form a **planetary nebula** — a balloon of hot gas like the famous Helix nebula in the constellation Aquarius (see Color Plate 4.1). The gas is **ionized** by ultraviolet light emitted by the surface of the star seen at the center of the Helix, which is very hot because what is now the central star used to be in the interior of the red giant. As electrons collide with positive ions and also recombine with them, the gas emits light and glows incandescently, often having the appearance of a ring or a disk in projection (hence the name "planetary nebula," but this formation has nothing to do with real **planets**). The nebula is enriched in helium, carbon, nitrogen (produced as a consequence of hydrogen burning at high temperatures; see Chapter 5), some oxygen, and trace quantities of heavier elements created by the capture of neutrons (see Chapter 5). The heavy elements subsequently become part of the **interstellar medium**, the gas and dust between the stars.

Besides the planetary nebula, all that remains of the original star is a small, dense, **white dwarf** seen at the center of the nebula. The white dwarf consists of a carbon-oxygen core and a helium shell, the products of nuclear burning that sustained the star's power output over billions of years. Our Sun will become a white dwarf in 6 to 7 billion years. Slightly more than half the Sun's current mass will be compressed into a sphere the size of Earth, at a **density** (mass/volume) so high that a tablespoon of the material will weigh several tons. It will be supported against the enormous pull of gravity by a purely **quantum-mechanical** pressure known as **electron degeneracy**. "Degenerate electrons" are not at all morally reprehensible; they just behave in a way that is highly unusual in comparison with classical low-density matter such as wood and bricks (see Chapter 5).

Having no new source of energy, the atomic nuclei in the white dwarf gradually cool down while supported by electron degeneracy pressure. The white dwarf will slowly fade and become a dark rock hurtling through space, essentially forever. Thus, despite a burst of glory in old age manifested by the red giant and planetary nebula phases, the Sun's death will be relatively quiescent. This is the case for most stars; indeed, all single stars initially less massive than about 8 $M_\odot$ end their lives as carbon-oxygen white dwarfs, and at least some in the range 8 to 10 $M_\odot$ probably become oxygen-neon-magnesium white dwarfs. The same is true of stars in widely spaced binary systems in which transfer of substantial amounts of gas from one star to the other cannot occur.

STELLAR EXPLOSIONS — CELESTIAL FIREWORKS

In some cases, stars can go out with a bang; they undergo explosions more violent than any except the **Big Bang** itself (see Chapter 1). Some **supernovae** brighten by a factor of 10^{12} to 10^{13}, as compared with 10^2 to 10^6 in an ordinary **nova**, which involves energetic processes (like nuclear fusion) only at the surface of a white dwarf. Supernovae are exciting and fun to watch (as are most bombs when not used for destructive purposes — who does not enjoy a good fireworks show?). For a few weeks, the visual brightness of a supernova can rival that of an entire **galaxy** containing 100 billion stars. The outer layers of gas are heated to temperatures of several hundred thousand K and are ejected at speeds up to one-tenth the speed of light ($0.1c$). A recent, well-studied example is SN 1987A, the first supernova discovered in 1987. This object (see Color Plate 4.2), located in the **Large Magellanic Cloud** (LMC) only 170,000 **light-years** away, was the brightest supernova to grace our skies since Kepler's supernova of 1604.

Supernovae are incredibly important and interesting objects. Compact remnants called **neutron stars**, in which a mass equal to that of 470,000 Earths ($1.4\ M_\odot$) is squeezed into a sphere whose diameter is less than that of a large city such as Los Angeles, are produced by some supernovae. This form of highly compressed matter cannot be created in laboratories on Earth, but nature has done it for us, which gives us the opportunity to indirectly study the properties of the compressed matter. Some supernovae may form **black holes** — regions in which matter is compressed so much that the local gravitational field permits nothing, not even light, to escape.

Supernovae send **shock waves** (pressure discontinuities produced by supersonic motion) through the interstellar medium, heat the tenuous gases to millions of K, and affect the structure of galaxies. Through compression, shock waves can also help initiate the formation of new stars in denser clouds of gas. Thus, the death of one star can trigger the birth of another star.

Supernovae are useful for **cosmology**, the study of the overall structure of the Universe. The tremendous power of supernovae makes them visible from far away, so they are attractive tools with which to measure distances between galaxies. The technique is analogous to the way the human brain judges the distance of an oncoming car by estimating the apparent brightness of the headlights and comparing with how bright they would appear if very nearby.

But perhaps the most important fact about supernovae, from the human perspective, is that they create and disperse most of the heavy elements, thus providing the necessary ingredients for Earth-like planets and life. Studies have shown that the Universe started out consisting only of hydrogen, helium, and trace amounts of lithium and beryllium; heavier elements were synthesized through nuclear reactions deep inside stars, and produced, as a by-product, **electromagnetic radiation** (light). If these elements remained forever locked up within the stars, they would be of no use. Explosions are generally necessary to release them, like most of the oxygen that you breathe, the phosphorus in your DNA, the calcium in your bones, and the silicon in rocks. Moreover, the explosions themselves directly or indirectly produce a majority of the heaviest elements, such as the iron in your red blood cells, the gold in your jewelry, the lead that shields your body in a dentist's X-ray machine, and the uranium used in nuclear reactors. (See Chapter 5 for a detailed discussion of the origin of different elements.)

The chemically enriched gases are ejected into the cosmos mostly via **supernova remnants** (see Color Plate 4.3) which gradually spread out, as shown in Color Plate 4.4. Over time, the gases merge and mix with the hydrogen and helium, from which our galaxy formed, as well as with the polluted debris from other supernovae and (to a lesser extent) novae, planetary nebulae, and stellar winds. When a sufficiently large quantity of gas collects in one cloud, it can begin to gravitationally contract, and stars form in the central regions (see Chapter 3). For example, we know that stars have recently been created in the Orion Nebula (see Color Plate 4.5 and Chapter 3, Color Plate 3.1). Some chemically enriched stars may have rocky, Earth-like planets on which life could eventually arise (see Chapter 6). In a similar way, our solar system formed 4.6 billion years ago out of gases that were chemically enriched, largely by many previous generations of supernovae. As Carl Sagan (1980) has said, "we are made of starstuff" — we owe our existence to supernovae.

HOW TO FIND A SUPERNOVA

Astronomers are very actively studying supernovae to learn more about their physical characteristics, explosion mechanisms, and consequences. However, there is a dearth of nearby objects; on average, supernovae occur only two or three times per century in large galaxies, such as the magnificent spiral M83 (see Color Plate 4.6), and some are hidden from our view by gas and dust in the interstellar medium. Thus, to have a reasonable chance of discovering several supernovae in any given year, one must look at many galaxies, most of which will be relatively far away because there are so few nearby galaxies. (It is like trying to find a two-headed snake; they exist, but are rare, so you are unlikely to see one in your own backyard.)

A careful amateur observer using a small telescope can make a valuable contribution to astronomy by searching for supernovae. The person can simply examine galaxies, like the one in Color Plate 4.6, as frequently as possible and look for any changes in appearance. The individual stars in Color Plate 4.6 are foreground stars in our own Milky Way galaxy, but an exploding star in the distant galaxy would have a similar appearance; thus, the observer should determine whether there are any additional stars in the galaxy each time it is viewed. On the evening of March 28, 1993, for example, Francisco Garcia discovered SN 1993J (the tenth, "Jth," supernova of 1993) in this manner with his 0.25-m (10-in.) reflecting telescope near Madrid, Spain. Only 12 million light-years away in the spiral galaxy M81, this was a very interesting object.

The champion of all visual supernova hunters is the Australian amateur astronomer Reverend Robert Evans; during the past 15 years he has discovered nearly 30 supernovae. He has memorized the star patterns around many hundreds of galaxies, and he can find and scan them very quickly, with a typical time of only 1 minute per galaxy. Occasionally, when he sees something suspicious, he makes a more detailed comparison by checking his printed chart of the galaxy. If the new object is still visible at the same location several hours later (decreasing the possibility of an asteroid), it becomes an excellent supernova candidate.

The same procedure can be used by amateur astronomers having small telescopes with attached cameras — new images of a galaxy can be compared with old images to find supernova candidates. This is especially easy to do with a charge-coupled device (**CCD**) rather than with a conventional photographic emulsion in the camera; CCDs are very sensitive and provide good images of faint stars, and the data can be manipulated digitally. On the night of April 1, 1994, several independent groups of amateur astronomers used CCD cameras to discover SN 1994I in the beautiful Whirlpool galaxy M51.

One can also conduct the search robotically by using a computer program to control the telescope and the CCD camera. My group at the University of California at Berkeley, for instance, has used a 0.8-m (30-in.) telescope at Leuschner Observatory (near the UC Berkeley campus) in this manner; no operator is present at the telescope during observations. The computer program even compares the new images of galaxies with old ones in search of supernovae. In less than 2 years of operation, we found seven supernovac, including SN 1994D in NGC 4526 (see Color Plate 4.7) 50 million light-years away in the Virgo cluster of galaxies.

News of a supernova discovery can be rapidly transmitted to the International Astronomical Union's Central Bureau for Astronomical Telegrams, which alerts appropriate astronomers to confirm the object and classify its type. (In 1995, at the time of this writing, the telephone number was 617-495-7244 or 7440, the fax number was 617-495-7231, and the e-mail address was marsden@cfa.harvard.edu or green@cfa.harvard.edu.)

SUPERNOVA CLASSIFICATION

After a supernova is discovered and confirmed, astronomers measure its **light curve** (brightness as a function of time; see Figure 4.8) and **spectrum** (brightness as a function of wavelength or color; see Figure 4.9), usually with CCDs. Both provide clues to the nature of the star and its explosion mechanism. Light curves are often measured through different filters transmitting blue, yellow, or red light, for example, because the temporal behavior of a supernova's brightness depends on wavelength. The spectrum is produced by using a prism or a **grating** (a piece of glass with many parallel grooves cut into its surface) to spread out the electromagnetic radiation into its component wavelengths from violet to red, much as raindrops turn sunlight into a rainbow. The spectrum can be used to determine the expansion velocity of the star, the chemical composition and temperature of the ejecta, and other parameters of interest.

Supernovae come in a variety of distinct types and subtypes. The primary classification system is based on the optical spectra, with light curves providing secondary information. If the spectrum does not exhibit any **absorption lines** (valleys) or **emission lines** (peaks) corresponding to hydrogen, the supernova is called Type I. If hydrogen is present, either in emission or absorption, the supernova is called Type II. Before classifying a supernova as Type II, however, one must make sure that the hydrogen emission is associated with the exploding star rather than with surrounding gas, as in a nebula like Orion (see Color Plate 4.5). This may be accomplished by determining whether the hydrogen lines appear broad.

Figure 4.9 illustrates typical spectra of Type I and Type II supernovae, roughly a week after maximum brightness. (Type I supernovae actually come in several varieties that have subtle differences, but we discuss here only the classic Type Ia objects that have a strong absorption line caused by singly ionized silicon in their spectra.) The lines are broad; gases are going both toward us and away from us and produce **Doppler shifts** in the detected light. The width of the lines tells us that gases move outwards at speeds exceeding 5000 km/s, and in some objects reach 30,000 km/s, which reflects the expansion of the supernova ejecta.

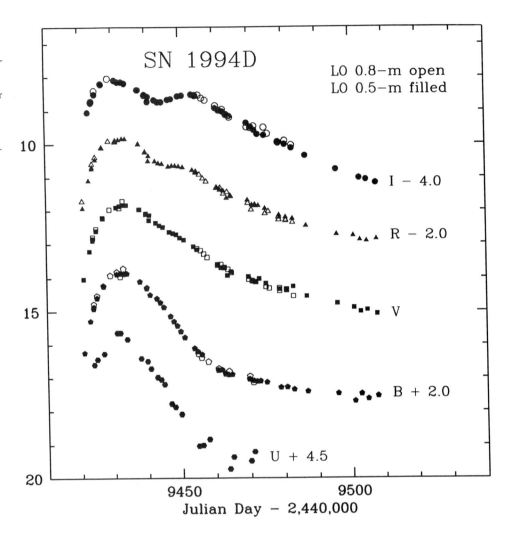

FIGURE 4.8

Light curves of SN 1994D through various filters, obtained by the author and collaborators with 0.5-m and 0.8-m robotic telescopes at Leuschner Observatory. *U*, near-ultraviolet; *B*, blue; *V*, visual; *R*, red; and *I*, near-infrared. The abscissa gives time in days. The ordinate units are **magnitudes**, a logarithmic measure of apparent brightness; a difference of 5 magnitudes corresponds to a brightness ratio of 100. The scale is "backwards": faint stars have larger magnitudes than bright ones. All light curves except the visual one have been offset for clarity.

This test — hydrogen or no hydrogen — is easy to perform; the **Hα** emission line of hydrogen (resulting from transitions between the third and second electron energy levels) is generally strong, and it falls in the red spectral region (~ 6500 Å) to which most CCDs are very sensitive. The test is also natural; hydrogen is by far the most abundant element in the Universe. Detailed analysis of the spectra of Type I supernovae demonstrates that hydrogen really is absent (or nearly absent) in the **progenitor** stars just before they explode, rather than somehow being hidden from view. Thus, it is reasonable to postulate that the stars giving rise to Type I and Type II supernovae are different and that their explosion mechanisms may also be distinct.

This classification scheme was proposed in 1941 by Rudolph Minkowski of the Mt. Wilson and Palomar Observatories. His colleague Fritz Zwicky, an avid supernova sleuth, discovered 122 supernovae during his lifetime — a record that is still unbeaten. Zwicky created three additional classes of supernovae (Types III, IV, and V), but there were only one or two examples of each, and they are now considered to be peculiar members of the Type II class.

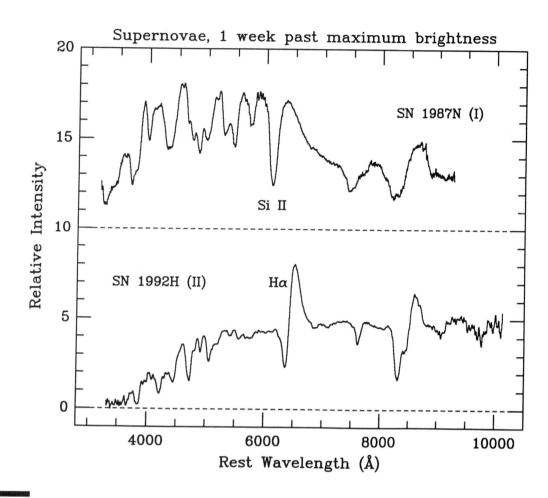

FIGURE 4.9

Optical spectra of supernovae, showing both a classic Type I (SN 1987N) and a Type II (SN 1992H). Spectra were obtained by the author and collaborators with the 3-meter Shane reflector at Lick Observatory.

Explosion Mechanism — Type I

Type I supernovae generally constitute a rather homogeneous class. At early times (within 1 month after the explosion), their spectra are dominated by lines of silicon, sulfur, calcium, oxygen, and magnesium. Although the detailed appearance of the spectrum gradually changes, at any given phase (time after explosion) the spectroscopic similarities among different Type I supernovae are striking. Indeed, through careful examination of the spectrum of a Type I supernova, it is possible to determine its phase rather accurately.

Type I supernovae are found in all kinds of galaxies, including ellipticals, which seem to contain primarily old stars (see Chapter 2). In spiral galaxies, the supernova rate tends to correlate with the rate of relatively recent star formation — but they show no clear preference for regions of current star formation, where the most massive stars live and die. This, together with the absence of hydrogen, suggests that we are generally witnessing the explosion of white dwarfs having ages of a few hundred million to a few billion years. The uniformity of the light curves and spectra of Type I supernovae further implies that the white dwarfs achieve a well-defined configuration before exploding.

The leading hypothesis is that a Type I supernova results from an uncontrolled chain of nuclear reactions in a carbon-oxygen white dwarf. This runaway is triggered when the white dwarf's mass gets close to 1.4 $M_\odot$, known as the **Chandrasekhar limit** after the Indian astrophysicist who first calculated it (see Chapter 5). Enough energy is released through nuclear fusion of light elements to blast the entire star apart; no compact remnant is created. Moreover, large quantities of **radioactive nuclei**, especially nickel, are produced through **nucleosynthesis**. The radioactive nickel subsequently decays into radioactive cobalt and finally into stable iron; this decay powers the optical light curve. (The radioactive nuclei emit gamma rays, most of which are trapped in the ejecta and are subsequently converted to optical and infrared photons.) Tycho's supernova of 1572, whose remnant is shown in Color Plate 4.3, appears to have been Type I.

The white dwarf probably grows toward the Chandrasekhar limit by **accretion** of matter from a companion star in a binary system (see Color Plate 4.10), although the details of the process remain a mystery. If the companion is a reasonably normal star donating hydrogen to the white dwarf, why do we not see hydrogen in the supernova's spectrum? Moreover, to avoid the surface explosions of novae (which generally lead to a decrease in mass) and other undesirable characteristics, the white dwarf's hydrogen accretion rate must lie within a fairly narrow range of values. Alternatively, the white dwarf could be accreting helium from a star that has lost its outer envelope of hydrogen, but such systems appear to be very rare. The white dwarf might also reach the Chandrasekhar limit by merging with another white dwarf; however, despite extensive searches, no binary white dwarfs with sufficiently small separations have been found in the Milky Way galaxy.

Explosion Mechanism — Type II

Type II supernovae are always found in spiral or irregular galaxies, usually in spiral arms and near nebulae, sites of active star formation. Thus, they are associated with massive stars, whose lives are so short that they have no chance to wander far from their birthplaces before exploding. (Although massive stars have more fuel to fuse, they do so at a prodigious rate, and use up the fuel much more quickly than do low-mass stars. For example, a 20 solar-mass star has a main-sequence lifetime of only about 10 million yr; that of the Sun is 10 billion yr.) The association is consistent with the observed heterogeneity of spectra and light curves of Type II supernovae; depending on their initial mass and chemical abundances, massive stars can have a wide range of sizes at the time of the explosion, and they can be embedded in gases (released by stellar winds) having vastly different densities.

Typically, the progenitors of Type II supernovae are thought to exceed 10 $M_\odot$ (some might be produced by stars in the range 8 to 10 $M_\odot$). In the cores of massive stars, hydrogen fuses to helium, helium to carbon and oxygen, and so on, to iron (see Chapter 5). The ashes of one set of nuclear reactions become the fuel for the next set, and an onion-like radial structure develops (see Color Plate 4.11). One pioneer in determining the detailed steps was Willy Fowler of the California Institute of Technology. (He recently died at the age of 83.)

The sequence of nuclear fusion stops at iron, the most tightly bound nucleus; fusion of iron into heavier elements does not liberate energy, it requires energy. Thus, an iron core builds up in the star, which by this time is generally a red **supergiant** up to 1000 times larger than the Sun (if our Sun were such a star, the orbit of Mars would be inside it). But the iron core cannot continue to grow indefinitely. When it reaches a mass comparable to the Chandrasekhar limit for white dwarfs, it is no longer able to support itself. The core begins to collapse, and electrons and protons combine to form neutrons and **neutrinos** (massless, or nearly massless, neutral particles; see Chapter 2). In the process, the core releases a tremendous amount of energy — effectively, the gravitational **binding energy** of the newly formed neutron star having a mass of 1 to 2 $M_\odot$ and a diameter of only 20 to 30 km. (Matter is compressed into such a small volume that its gravitational mass is actually less, by ~ 0.1 $M_\odot$, than when it was spread farther apart. This is analogous to the mass deficit of tightly bound atomic nuclei.) Most neutrinos, being very noninteracting particles, escape almost immediately.

When its density exceeds that of atomic nuclei, the central region rebounds like a basketball dropped to the floor. As it smashes into the surrounding material, it creates an outward moving shock wave that reverses the collapse and leads to an energetic expansion. A similar bounce effect can be seen if a tennis ball is placed atop a basketball and both are dropped simultaneously. The copious neutrinos and antineutrinos (the **antiparticles** of neutrinos) emitted by the exceedingly hot (100 billion K) neutron star probably help eject the layers of gas surrounding the core; only about 1% of their energy needs to be transferred to the ejecta. During the explosion, heavy elements such as nickel, zinc, and platinum are synthesized through nuclear reactions, although the relative proportions of the elements differ from those in Type I supernovae.

The basic idea of core collapse in a massive progenitor, followed by rebound, was formulated by Fritz Zwicky and his colleague Walter Baade at Caltech in the 1930s. Although the neutron had been discovered by James Chadwick only 1 year earlier, Zwicky and Baade boldly predicted that stars made almost entirely of neutrons are created by this mechanism. The density of the material would be close to that of nuclear matter — a tablespoon of it would weigh about 1 billion tons.

SUPERNOVA 1987A — A GIFT FROM THE HEAVENS

By the early to mid-1980s, detailed theoretical studies of supernovae were being done with complex numerical codes running on large computers. Insights were gained into the nature of the progenitors, the explosion mechanisms, and explosive nucleosynthesis for both Type I and Type II supernovae. The predicted light curves and spectra generally agreed with observations, at least in broad terms, but other problems remained; for example, the bounce mechanism after core collapse in a Type II supernova could not be understood reliably, and there was no consensus on how a white dwarf actually reaches an explosive configuration to form a Type I supernova.

What was needed were some bright, nearby supernovae to test and refine the theories. Extensive observations throughout the electromagnetic spectrum are important because radiation at different wavelengths provides different clues to the phenomenon. Ideally, a supernova within a few thousand light-years could be found in the Milky Way galaxy; after all, our galaxy should produce at least two supernovae per century. However, most of these supernovae are distant and are obscured by patchy intervening clouds of gas and dust (galactic smog) (see Color Plate 4.12) because the Sun sits in the plane of the Milky Way galaxy. The most recent easily visible supernova was observed by Johannes Kepler in 1604. One was seen in 1572 by his mentor, Tycho Brahe, and a more recent supernova (Cassiopeia A) occurred around 1680 but was invisible (or only barely visible) to the naked eye (see Chapter 5, Figure 5.5).

If a nearby supernova in our galaxy is too much to hope for, the next best alternatives are to look at the LMC and the Small Magellanic Cloud (see Color Plate 4.12), two dwarf satellite galaxies of the Milky Way about 170,000 and 210,000 light-years away, respectively. The LMC, in particular, is home to a gigantic stellar nursery known as the Tarantula nebula (30 Doradus; see Color Plate 4.2); many massive stars have formed in this region within the past 100 million years, which makes the LMC an excellent potential site for a supernova.

In 1987 our dream came true — a supernova (SN 1987A) near the Tarantula nebula was discovered by Ian Shelton at Las Campanas Observatory in Chile (operated by the Carnegie Institution of Washington) and, independently, a few hours later by several other observers in New Zealand and Australia. Shelton, an operator for the University of Toronto's 0.6-m telescope, had initiated a search for **variable stars** in the LMC with a spare 0.25-m refracting telescope; on February 23, 1987, he took his first good photograph of the LMC. On February 24 he obtained his second photograph — but the wind blew shut the roll-off roof of the telescope shed, so he decided to develop the image immediately, before the night ended. Comparing the two photographs, he noticed a bright new star in the image obtained on February 24 and confirmed its presence by looking at the LMC.

Color Plate 4.13 shows "before and after" photographs of SN 1987A. It is not difficult to imagine the excitement of astronomers; we had been waiting nearly 4 centuries for such an event. The drama was conveyed to the general public — SN 1987A was the cover story for *Time* magazine (March 23, 1987, issue), major newspapers, and amateur astronomy magazines such as *Sky & Telescope* (May 1987 issue).

Word of the discovery spread fast; soon most optical and radio observatories in the southern hemisphere (from which the views of the LMC are best) devoted large amounts of time to SN 1987A. To collect X-ray and gamma-ray **photons**, to which Earth's atmosphere is largely opaque, rockets and balloons containing the appropriate detectors were launched to high altitudes. Ultraviolet photons, which are blocked by Earth's ozone layer, were observed with the orbiting International Ultraviolet Explorer. Good data at infrared wavelengths were obtained with the Kuiper Airborne Observatory; at an altitude of more than 40,000 feet, this flying Observatory is above most obscuring water vapor.

Testing the Theories

Optical spectra of SN 1987A revealed the presence of hydrogen, which made it a Type II supernova. A major test of the theoretical models, then, was to see whether the progenitor was a massive, evolved star. Careful examination of existing photographs of the LMC, obtained before the explosion, verified that the progenitor, known as Sanduleak −69°202, was a supergiant with a probable initial mass of 18 to 20 $M_\odot$ and an age of roughly 11 million years.

There was a twist, though — the star was a blue supergiant, hotter and smaller than a red supergiant. This supergiant is now thought to be a consequence of the low abundance of heavy elements in the LMC (a dwarf galaxy) in comparison with the Milky Way (a much more massive galaxy); the structure of the atmosphere of a star depends on its **opacity**, or degree of transparency, which is related to its heavy-element content. Conditions in the deep interior of a blue supergiant are similar to those in a red supergiant, so the explosion mechanism should have been essentially the same. The relatively small size of the progenitor was also consistent with the peculiar light curves of SN 1987A — the object appeared unexpectedly faint because considerable energy was required to expand the star. Thus, SN 1987A confirmed that Type II supernovae arise from massive stars but also provided valuable new information on the possible light curves and progenitor properties.

Another crucial prediction of the theory is that new elements are produced through explosive nucleosynthesis. Among the most important of these is a radioactive **isotope** of nickel, which decays with a **half-life** of a week to radioactive cobalt and subsequently (half-life, 2.5 month) to stable iron. The radioactive nuclei are in excited states that emit gamma rays as they drop down to lower energy levels, in much the same way that an excited electron in an atom emits visible or infrared light when it jumps to a lower level. Moreover, as in the case of atoms, each nucleus produces a unique spectral pattern of photons. Thus, one way to confirm the synthesis of heavy elements is to search for the spectral signature of radioactive nickel or cobalt.

This was done with gamma-ray telescopes in the Solar Maximum Mission satellite and in gondolas attached to large balloons flown high over Australia and the Antarctic. Six to 12 months after the discovery of SN 1987A, photons having precisely the energies corresponding to radioactive cobalt (as measured in terrestrial laboratories) were detected and confirmed. (Delay was caused by the opacity of the supernova to gamma-ray photons; the ejecta had to expand and thin out before a few of the gamma rays could escape from the star and travel unimpeded toward Earth, 170,000 light-years away.) The gamma rays had to have been emitted by nuclei synthesized during the explosion; if the cobalt were instead produced much earlier (e.g., in the gas that formed the star), it would have long ago decayed to stable iron.

The results were bolstered by infrared spectra, which showed an overabundance of nickel and cobalt, and by the optical plus infrared light curve, whose rate of decline 4 to 16 months after the explosion matched the radioactive decay rate of cobalt. (During this interval, most gamma rays remain trapped in the ejecta and convert their energy to optical and infrared photons, which escape immediately. Thus, the optical plus infrared brightness is proportional to the gamma-ray luminosity, which itself depends on the amount of remaining cobalt.) Measurements show that about 0.07 $M_\odot$ of radioactive nickel were produced by the explosion. Such an unequivocal confirmation of explosive nucleosynthesis was a major breakthrough and demonstrated beyond reasonable doubt that heavy elements are indeed cooked and dispersed by supernovae.

Neutrinos from Hell

Perhaps the most spectacular result provided by SN 1987A was the direct evidence for the formation of a neutron star. This measurement was fortuitous; it was made by separate experiments in Japan by Kamiokande and in the United States by the Irvine-Michigan-Brookhaven collaboration (IMB). The experiments were designed to test certain **grand unified theories** of physics that predict that the proton is slightly unstable with a half-life exceeding 10^{30} years. The apparatus consists of an underground tank containing several thousand tons of ultrapure water and surrounded by detectors sensitive to visible light. Because the speed of light is only about $0.7c$ in water, slower than its speed in a vacuum ($1.0c$), particles can exceed the local speed of light without violating the known laws of physics. If a charged particle is sent through the water at a speed faster than the local speed of light, it emits a cone of blue **Cherenkov radiation**, the electromagnetic equivalent of the sonic boom heard when an airplane exceeds the speed of sound. This radiation, along with its arrival time, can be recorded and analyzed.

How is this relevant to SN 1987A? A young neutron star is expected to be very hot — around 100 billion K. At such high temperatures, energy escapes from the star primarily in the form of neutrinos and antineutrinos. Despite not being inclined to interact with matter, a very small fraction of neutrinos and antineutrinos do interact. This is what happened in the Kamiokande and IMB water tanks — antineutrinos combined with protons to form neutrons and high-energy **positrons** (antielectrons), and the positrons sped through the water faster than the local, depressed speed of light and thereby produced Cherenkov radiation that was subsequently detected. (Another possibility is that a neutrino kicked an electron to a very high energy, but this event is rare compared with the antineutrino/proton interaction.)

The Kamiokande and IMB teams each detected Cherenkov light from about 10 positrons, in agreement with expectations; roughly 30 billion neutrinos and antineutrinos passed through every square centimeter of Earth when the "flash" arrived, but their interaction probability is exceedingly small. (Indeed, only one person in a few thousand experienced a direct interaction.) Although 10 positrons may seem insignificant, it was a clear indication that a neutron star had been produced by SN 1987A — the antineutrinos arrived at the correct time (based on an extrapolation of the optical light curve to the moment of the explosion) and with the expected energies. This monumental discovery marked the birth of *extrasolar neutrino astrophysics*; previously, the only cosmic neutrinos ever detected had been from the core of the Sun.

The energy released by the core collapse of SN 1987A exceeded 10^{53} **ergs** — about $0.1\,M_\odot$ of material converted into pure energy according to $E = mc^2$. This is close to the energy produced in 1 second by the sum total of normal stars in the observable part of the Universe! Most (99%) of the energy of SN 1987A was carried away from the neutron star by neutrinos and antineutrinos within a few seconds after the core collapsed. About 1% was the kinetic energy of the ejecta gained largely from interactions with these elusive particles, and less than 0.01% came out at visible and infrared wavelengths. Despite being visually spectacular objects, Type II supernovae are, fundamentally, giant generators of neutrinos and antineutrinos.

SN 1987A provided an interesting upper limit to the **rest mass** of the electron neutrino. (Two other types are the muon and tau neutrinos; also, particles and antiparticles have the same rest mass.) If the neutrino were massless, it would be forced to travel at the speed of light, in which case all the neutrinos should have arrived simultaneously if they were emitted instantaneously. If, on the other hand, neutrinos have mass, those with higher energies move faster and arrive earlier than the low-energy neutrinos. In fact, the antineutrinos arrived within a time interval of about 10 seconds, which suggests that they have a mass of about 15 **electron volts** (eV). (For comparison, in these units the mass of an electron is 511,000 eV.) However, the antineutrinos were probably not all emitted instantaneously; the hot, newly formed neutron star is opaque to neutrinos, and it takes time for them to leak out. The exact calculation is difficult, but the entire observed spread in arrival times could conceivably be attributed to this effect. Thus, the mass 15 eV should be viewed as an upper limit. This mass is consistent with the best existing upper limits (based on laboratory measurements) to the mass of the electron neutrino.

Mainly as a result of SN 1987A, we now have a reasonably good understanding of the explosion mechanism of Type II supernovae. Recent two-dimensional calculations show how matter heated by neutrino interactions is able to move outward through **convection**, which greatly increases the ease with which a successful explosion can be simulated. Abundances of elements synthesized during the explosion, together with their distribution in the ejecta, are also being determined by observations and theoretical work. Although progress has been made on detailed calculations of white dwarfs undergoing nuclear runaway, we have yet to test the theory of Type I supernovae as extensively; a nearby example that can be studied with a wide variety of telescopes is needed.

NEUTRON STARS

As mentioned previously, Zwicky and Baade predicted in the 1930s that supernovae should produce neutron stars. (Some neutron stars might also form by the collapse of white dwarfs accreting matter from a companion star; these white dwarfs escape death as Type I supernovae for some values of accretion rate and initial mass.) Neutron stars support themselves by **neutron degeneracy** pressure, similar to the electron degeneracy pressure of white dwarfs. A few physicists, such as J. R. Oppenheimer and G. Volkoff, did theoretical studies of the possible properties of such stars, but there were no observations with which to compare. A neutron star is only about 20 to 30 km in diameter, which makes it difficult to see directly, at least in terms of radiation emitted uniformly from its surface. Very young (hot) neutron stars or neutron stars accreting matter from a bound companion star can emit X-rays, but X-ray telescopes were not in use until the 1960s and 1970s. Over 4 decades passed before neutron stars were actually detected — and the discovery was accidental, much like that of neutrinos from SN 1987A.

In 1967, Jocelyn Bell (now Burnell) and her doctoral thesis advisor, Antony Hewish, were using a radio telescope in Cambridge, England, to study short-timescale variations in the apparent radio brightness of astronomical objects. These are generally produced by inhomogeneities in the density of interplanetary ionized gas through which the radiation passes. Much to her surprise, Bell found that one part of the sky seems to emit very regularly spaced pulses of radio waves. Their intensity varied considerably, but the time interval between pulses was always 1.3373011 seconds.

No known astronomical objects produced such regular, rapid pulses of radiation. After eliminating possible terrestrial sources, the astronomers briefly considered the idea of communication signals from intelligent extraterrestrial life (indeed, the object was sometimes half-jokingly called an LGM, for "little green men"). Bell, Hewish, and their collaborators, however, soon discovered three additional objects having the same signature, but pulse periods of 0.253065, 1.187911, and 1.2737635 seconds. Other similar objects (but having different periods) were subsequently found; an example of such a **pulsar** is shown in Figure 4.14. It was unlikely that different intelligent civilizations, in widely separated regions of the Milky Way galaxy, used exactly the same method by which to communicate. Moreover, if the signal from one of the pulsars were coming

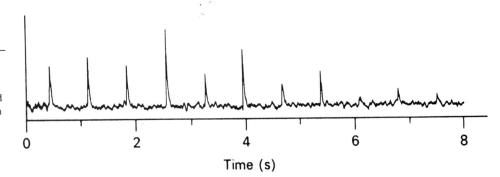

FIGURE 4.14

Chart record of individual pulses from PSR 0329+54, one of the first pulsars discovered. The pulse period is 0.7145 seconds. (Courtesy Joseph H. Taylor.)

from a planet, a slight periodic shift in the pulse arrival times would be expected as a result of the planet's orbital motion, but none was detected. A different explanation had to be found.

The pulses could not be coming from oscillations in size of normal stars, such as in the case of **Cepheid variables**, since the natural periods of stars are much longer than 1 second. (The Sun, for example, experiences low-amplitude oscillations with a period ~ 5 min.) Moreover, light from the limb (visible edge) of a star would have a longer distance to travel than light from the near side (about a 2-s delay for the Sun), which would thereby smear out the pulses — yet they were observed to be very narrow, with a width of only about 5% of the period. The natural oscillation period of a white dwarf, which is denser than a normal star, is 1 to 10 seconds — not quite fast enough for the most rapid normal pulsars (0.03–0.1 s). Conversely, that of a neutron star is ~ 0.001 seconds, which is too fast. Stellar oscillations of any sort therefore seemed unlikely.

Perhaps two stars were orbiting each other with a period of roughly 1 second. This clearly is not the case with normal stars because they cannot get sufficiently close together. (**Kepler's third law** of orbital motion states that the square of the period of revolution is proportional to the cube of the major axis of the elliptical orbit; scaling from the Earth-Sun system, one finds that the stars would have to be inside each other.) The lowest orbital period of two white dwarfs is about 2 seconds, not short enough. Two neutron stars, or a white dwarf plus neutron star pair, do not have this problem. However, the separation between the stars would be so small that they would emit **gravitational waves** (ripples in the curvature of **space-time**) as they orbit each other, according to Einstein's **general theory of relativity**. The system would lose energy, and the two stars would get even closer together, which would lead to a detectable shortening of the period. Pulsars, however, were nearly perfect "clocks"; their periods seemed stable. Subsequent, more precise measurements showed that the periods were very gradually increasing, but certainly were not decreasing.

Having eliminated oscillation and orbital motion, astronomers were left with the possibility of rotation. A normal star is out of the question; its surface would have to exceed the speed of light if it were rotating once per second. A dense white dwarf is stable if rotating more slowly than a few times per second, but at shorter periods it would be torn apart by **centrifugal forces**. A rotating neutron star, on the other hand, does not have this problem; neutron stars are so dense that they can withstand rates to about 1000 rotations per second.

Thus, astronomers concluded that pulsars are probably rotating neutron stars. If a beam of radiation were to somehow emanate from the neutron star, along an axis not coincident with the rotation axis, we would observe "pulses" when the beam intersects our line of sight once per rotation period (or perhaps twice, if there were two oppositely directed beams whose axis is nearly perpendicular to the rotation axis). The effect would be analogous to that of a lighthouse: it is always "on," but we see it only when its beam crosses our line of sight.

What produces the beam? If neutron stars are highly magnetized, with a magnetic field pattern similar to that of Earth (a **dipole field**, which can be visualized by inserting the ends of many wires into the opposite poles of a ball), the rapid rotation induces an electric field according to the equations of electromagnetism. This will accelerate charged particles, such as electrons and positrons, predominantly along the poles because charged particles do not easily cross magnetic field lines. They, in turn, will radiate energy along their direction of motion, which results in a beam (see Figure 4.15). Some radiated photons interact with the magnetic field and are converted into electron-positron pairs and thereby escalate the process. Although the details are complicated and controversial, many astronomers believe that pulsars "shine" because of this mechanism or one of its variants.

The rapid rotation of a neutron star is easily understood. If the progenitor star rotated appreciably (as most well-observed stars seem to do), the neutron star would naturally attain a high rotation rate as a consequence of core collapse. Objects tend to retain their **angular momentum**, a measure of the amount of spin determined by the product of rotation rate, mass, and mass distribution. A large spinning star would have a much shorter rotation period after it shrinks, just as an ice dancer spins faster as she brings her arms closer to her body. Similarly, the magnetic field permeating a star may grow by large factors when the core collapses; the field strength is inversely proportional to the cross-sectional area of the star.

Observational Evidence

The energy emitted by a pulsar has to come from somewhere; conservation of mass plus energy is one of the fundamental laws of physics. In fact, the energy is produced at the expense of the neutron star's rotational energy; thus, the observed rotation periods of pulsars gradually increase. Eventually the rotation is so slow that the induced electric field is too weak to support the beam-generation mechanism, and the neutron star stops shining. (Some astronomers also believe that the magnetic field decays very slowly with time and thereby further contributes to the demise of the pulsar.) The process might take about 10 million years for a typical pulsar — far longer than the lifetime of most supernova remnants, whose gases completely merge with the interstellar medium within about 100 thousand years after the explosion. Hence, it is no surprise that most pulsars are not observed to be associated with known supernova remnants.

A few key examples, however, reinforce this scenario. Specifically, there is a rapidly spinning pulsar ($P = 0.089$ s) within the Vela nebula (see Color Plate 4.4), a supernova remnant about 20,000 years old. Even more striking is the pulsar near the center of the Crab nebula (see Color Plate 4.16), the expanding remnant of a supernova that was first seen on July 4, 1054 A.D. (Although observed extensively by Asian astronomers, this supernova appears to have escaped attention throughout most of Europe and North America.) This is the youngest known pulsar and, perhaps not coincidentally, it has the most rapid rotation (30 times per s) of any normal pulsar. Although the majority of pulsars emit almost entirely at radio wavelengths, the Crab pulsar is so young that it can

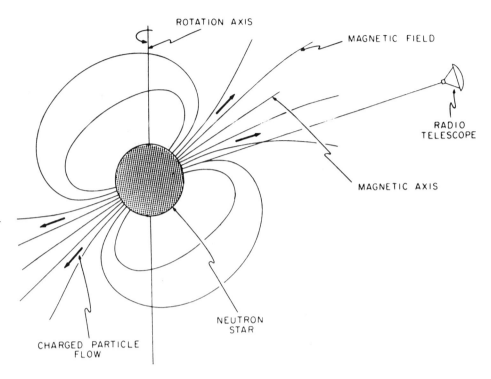

FIGURE 4.15

A schematic drawing of a pulsar showing the relationship between the neutron star's rotation and magnetic axes. A cone of radio emission is centered on the magnetic axis. Once per rotation, the beam sweeps across Earth (represented by a radio telescope) and we observe a pulse of radio waves. This is analogous to the lighthouse mechanism. (Diagram by David J. Helfand. Reproduced with permission of the Astronomical Society of the Pacific.)

easily be seen optically (see Color Plate 4.17). Having short periods and being associated with obvious supernova remnants, both the Vela and Crab pulsars played critical roles in the development of our understanding of pulsars.

The detection of antineutrinos provided convincing, but indirect, evidence that SN 1987A produced a neutron star, at least temporarily; a very hot surface had to exist from which 0.1 $M_\odot$ of gravitational binding energy was radiated in the form of neutrinos and antineutrinos. Nevertheless, to directly detect the neutron star would be exciting and important. After all, the neutron star may have subsequently accreted enough matter to become unstable and collapse to a black hole.

Searches for optical pulsations from SN 1987A have yielded one or two false alarms, but nothing definitive. Very little radio radiation is currently being emitted by SN 1987A. Moreover, the recent optical photographs from the Hubble Space Telescope (HST) do not show anything unusual at the center of SN 1987A. (They do, however, reveal two large rings and one smaller ring of gas that were probably produced by a wind emanating from the star before its explosion [see Color Plate 4.18]. These provide clues to the evolutionary history of the star.) One complication is that the ejecta of SN 1987A are still rather opaque at optical and radio wavelengths, which obscures our view toward the center. Nevertheless, if a powerful pulsar were inside, it would heat the surrounding material and cause it to glow much more brightly than what is observed. We can already rule out a source of energy in the center of SN 1987A exceeding about 0.01 of the luminosity of the Crab pulsar — yet SN 1987A is far younger than the Crab pulsar, so it should be much brighter if it were born with the same rotation rate and magnetic field strength. The neutron star could also liberate substantial amounts of energy through accretion of material from its surroundings. Perhaps the neutron star in SN 1987A actually has an unexpectedly long rotational period, or there is not much gas to accrete. Although these are valid possibilities, an alternative is that the center of SN 1987A now contains a black hole.

Millisecond and Binary Pulsars

There are two exotic subclasses of pulsars. First are **millisecond pulsars**: these objects rotate 50 to 1000 times/s, rather than 0.1 to 10 times/s as do most normal pulsars. If converted to audible sounds, the frequencies of millisecond pulsars would be similar to those of familiar notes near middle C of the musical scale. For example, PSR 1937+21 has a frequency of 642 Hz, close to E-flat in the treble clef; that of PSR 1953+29 is 163 Hz, roughly E in the bass clef. Dozens of millisecond pulsars are now known. These objects are thought to be old neutron stars (over 100 million yr old) spun up by accretion of matter from a companion star. Their emission mechanism has been rejuvenated by the very rapid rotation, despite a magnetic field that is 100 to 10,000 times weaker than that of normal pulsars.

Millisecond pulsars are incredibly stable clocks; their periods hardly change, and they pulse so many times per second that small deviations from regularity can be discerned over relatively short time intervals. This led to the discovery of tiny, cyclical variations in the pulse period of one of them. The obvious conclusion was that this particular pulsar is in orbit around at least one other object. Quantitative analysis of the data supports this hypothesis and suggests that there are three companions whose masses resemble those of planets. These planets, however, cannot be "conventional," that is, formed from the same gas and at the same time as the parent star; had they existed before the supernova that produced the neutron star, they would have been expelled from the system. Instead, they probably formed after the explosion, perhaps from debris remaining in the vicinity of the neutron star. It is also possible that they are the shattered remains of the companion star. Although not normal planets like Earth, these objects are very important; they show that planet-sized objects can form elsewhere outside our solar system, even in what initially appear to be environments adverse to their formation.

A few pulsars are known to be in binary systems with another neutron star. The first and best known of these **binary pulsars**, discovered by graduate student Russell Hulse and his thesis advisor, Joseph H. Taylor, has an orbital period of slightly less than 8 hours — the stars are so close together that they would nearly fit inside our Sun. Careful analysis of the pulse arrival times during more than a decade has shown that the orbital period is decreasing; the two stars are gradually spiraling toward each other. The observed rate is exactly equal to that predicted by the general theory of relativity; the system is losing energy through the emission of gravitational waves. This provides the best test of general relativity in reasonably strong gravitational fields. The importance of the discovery was officially recognized through the award of the 1993 Nobel Prize in physics to Hulse and Taylor. (Interestingly, Antony Hewish received the 1974 Nobel Prize in physics for the discovery of pulsars, but Jocelyn Bell did not. Hewish shared the prize with Sir Martin Ryle, one of the pioneers of radio astronomy.)

BLACK HOLES

The measured masses of most neutron stars in binary systems are close to 1.4 $M_\odot$. Above a certain mass, neutron degeneracy pressure should be unable to support a neutron star against gravitational collapse. The precise value of this limit is not well known, but is thought to be in the range 2 to 3 $M_\odot$. Very rapid rotation (say, 1000 times/s) might increase the maximum permissible mass of a neutron star to about

$4\,M_\odot$. Calculations suggest that some stars with large initial masses (10–100 $M_\odot$) have final masses that exceed this limit. They cannot lose enough matter through winds and the supernova explosion itself or through mass transfer in a binary system, so they will not end their lives as neutron stars. Similarly, a neutron star in a binary system could accrete enough material from its companion to be driven over the stability limit. In both cases, the resulting object is called a **black hole**.

A black hole is appropriately named — its gravitational field is so strong that *nothing*, not even light, can escape. To obtain a qualitative understanding of how this might be, consider Newton's law of universal gravitation ($F = GM_1M_2/r^2$, where G is the gravitational constant) applied to a ball of mass M_2 on the surface of Earth of mass M_1, a distance r (the radius) from the center of Earth. The force of gravity holds the ball to Earth's surface, and the ball must be thrown with a certain minimum initial speed, the **escape velocity** (about 11 km/s, neglecting air resistance), to fly completely away from Earth. Now, suppose the Earth were compressed to a sphere having one-half its current radius, while retaining all its mass. The gravitational force on the ball at the surface would be four times larger than before, and the escape velocity would be multiplied by the square root of 2, which corresponds to a new value of about 16 km/s. (The formula for the escape velocity is $v = (2GM_1/r)^{1/2}$.) If the Earth were further compressed to a sphere having one-fourth its true radius while retaining all of its mass, the gravitational force on the surface ball would grow by a factor of 16 and the escape velocity would increase by a factor of 2 (to 22 km/s).

One can imagine this process progressing to such an extent that the escape velocity formally becomes equal to the speed of light, in which case neither the ball nor anything else (including light) can escape and the object appears black. Indeed, such arguments were made as far back as 1783 by John Michell and 1795 by Pierre Simon de Laplace. Although the Newtonian version of mechanics (including the law of universal gravitation) is not valid when the gravitational field becomes very large, and Einstein's general theory of relativity must instead be used, the formula relating the minimum radius to which a nonrotating object of mass M must be compressed to form a black hole is the same in both cases: $R = 2GM/c^2$. Earth's radius, for example, would have to be no larger than 0.89 cm to form a black hole; that of the Sun would be about 3 km.

The radius of a nonrotating black hole, as defined above, is known as the **Schwarzschild radius**, after Karl Schwarzschild, who in 1916 used the newly developed relativity theory to formally derive the radius. Dead stars whose mass exceeds the maximum possible mass of a neutron star must continue collapsing until they are smaller than their Schwarzschild radius; thus, they form black holes. In Einstein's theory, gravitation is actually a curvature of space-time produced by any mass or any energy (according to $E = mc^2$), and the orbits of objects are simply their natural paths in this curved geometry. In the vicinity of a black hole, space-time is curved so highly that no possible trajectories lead out from within the Schwarzschild radius. In a sense, the black hole is a "pinched off" part of the Universe from which no information can flow to the outside world. The imaginary surface that separates the black hole from the rest of the Universe is known as the **event horizon**. This boundary has a radius equal to the Schwarzschild radius if the black hole is not rotating, but the radius can be as much as a factor of 2 smaller in a rapidly rotating black hole.

What is the fate of the collapsing star after its radius reaches the Schwarzschild radius? According to the equations of classical general relativity, the matter keeps falling to progressively smaller radii until it is technically a point of zero volume and infinite density known as a **singularity**. In other physical situations, however, the classical laws of physics break down when exceedingly small volumes are considered; rather, one must use the laws of quantum mechanics. These have been thoroughly tested in many ways, such as by predicting and measuring the energy levels of the hydrogen atom. The quantum world appears to avoid singular points and their associated infinite quantities; the structure of matter is instead described by probability distributions of nonzero spatial extent. Although a fully self-consistent quantum theory of gravity has not yet been developed, most researchers believe that a successful one will show that the singularity in a black hole is not really a point of infinite density. Nevertheless, the singularity is probably very small and dense — certainly any material object would be crushed beyond recognition within it.

Fun Facts about Black Holes

Black holes have many fascinating properties. For example, if you were to fall feet-first into a black hole, you would be stretched along the length of your body and squeezed along the width by the hole's **tidal forces**. The stretching occurs because the gravitational pull on your feet (which are closest to the black hole) significantly exceeds that on your head, and the difference increases rapidly as you approach the hole. Similarly, the squeezing is produced by the fact that all points are pulled toward the center of the black hole along radial lines; your two shoulders therefore get progressively closer together. Well before actually reaching the singularity, you would resemble a long, thin string of rubber that is being stretched from both ends.

The strength of the tidal forces is a function of the mass of the black hole. Black holes formed from individual stars have enormous tidal forces even outside the event horizon, but billion-solar-mass black holes such as those at the centers of some galaxies (see the next section) seem so benign that you would not feel anything unusual outside the event horizon. You could fall into a black hole without initially knowing that anything was amiss — but you would inexorably be drawn towards the singularity. Similarly, the average density of a black hole (defined as the hole's mass divided by the volume enclosed by the event horizon) is proportional to the inverse square of its mass — low-mass black holes have high average densities; gigantic black holes have low average densities. The singularity at the center of the black hole is believed in all cases to be extremely dense.

Another important effect is **time dilation** in the vicinity of a black hole. If you were far from a black hole and watching a friend fall in, your friend's clock would appear to run progressively more slowly as he approached the event horizon. From your perspective, time would be slowing down for your friend. Indeed, as he gets infinitesimally close to the event horizon, time slows to a halt; you never actually see him reach the event horizon because this takes an infinite amount of time from your point of view. It takes a finite (and short) time from your friend's perspective, however (this is not a method for increasing one's longevity!). On the other hand, if your friend were to approach the event horizon (always remaining outside it) and subsequently escape from the vicinity of the black hole by the appropriate use of rockets, he would have aged less than you did. Hence, this is a method for jumping into the future while aging very little, much like what happens when one travels at speeds close to that of light, according to the theory of relativity.

A related phenomenon is the **redshifting** of radiation emitted from the vicinity of (but outside) a black hole. In the above example, suppose your friend were emitting flashes of blue light once per second on his clock while he fell towards the black hole. Not only would you see the flashes arrive at progressively longer intervals (the result of time dilation), but the flashes would appear redder and redder — that is, the wavelength of light would be stretched. Because the energy of each photon is inversely proportional to its wavelength, photons lose energy as they climb out of the deep gravitational field surrounding a black hole. A photon that is attempting to escape from the event horizon itself is redshifted to infinite wavelength (zero energy); hence, we cannot detect it.

There is a famous theorem stating that "black holes have no hair." The gist is that black holes are very simple objects, completely described from the perspective of an outside observer by only three quantities: mass, electric charge, and angular momentum. In other words, external observations of a black hole cannot reveal the identity of objects that might have been thrown into it. Any small perturbations in the event horizon produced by an object falling into a black hole are quickly erased, which leaves only the three global properties of the hole.

Classically, the mass of a black hole can never decrease, because nothing can escape from within the event horizon. We expect the mass of a black hole to grow as objects in its vicinity are swallowed. However, Stephen Hawking showed that extremely small black holes do, in fact, evaporate at an appreciable rate as a result of a quantum-mechanical process. The lower the mass of the black hole, the greater its rate of evaporation — thus, low-mass black holes actually explode with a burst of high-energy radiation when their mass approaches zero. If tiny black holes were produced during the Big Bang, for example, the ones having initial masses comparable to those of big mountains (such as Mt. Everest) on Earth would now be exploding. (Less massive ones exploded earlier.) Initially, the occasional bursts of gamma rays detected in the sky by gamma-ray telescopes were thought by some to be these exploding **primordial** black holes, but that possibility has been eliminated because the observed properties of the bursts are inconsistent with theoretical predictions.

Detecting Black Holes

Do we have any concrete evidence that such bizarre creatures as black holes really exist in nature? Indeed we do, and it comes primarily from two types of objects.

The first is the category of **binary X-ray sources**. In the 1960s and 1970s, after X-ray satellites were launched above most of Earth's atmosphere, astronomers noticed that certain parts of the sky emit X-rays profusely. Close examination revealed that in one well-observed case (Cygnus X-1), the source of the X-rays appeared to be a bright star. If that star were in orbit around a neutron star or black hole companion, and the companion were accreting gas from the star (as in Color Plate 4.10), the falling gas would heat up and emit X-rays. (High-energy radiation is produced because the gas is greatly accelerated by the strong gravitational field.) Spectroscopic studies of the motion of the bright star showed that it is, in fact, bound to a companion having a mass of at least 7 $M_\odot$, and a probable mass of 16 $M_\odot$. If the companion were a normal star having this mass, it would be luminous and easily visible, yet no object is seen. The companion could not be a low-luminosity white dwarf or a neutron star because its mass greatly exceeds both the Chandrasekhar limit and the limiting mass of a neutron star. The most reasonable conclusion is that the companion is a black hole.

Similar systems were subsequently found. Although many turned out to contain neutron stars (with large quantities of energy released as matter hits the stellar surface), there are now a half-dozen well-established (and about a dozen possible) black hole candidates. The best one is a star known as V404 Cygni, an **X-ray nova** that erupted in 1989. The mass of the invisible companion is ~ 12 $M_\odot$, and there are fewer uncertainties than there were in the analysis of Cygnus X-1. Recently, my students and I measured the mass of the dark companion in another X-ray nova known as GS 2000+25 and found it to be at least 5 $M_\odot$. This is a very probable black hole; the mass exceeds the limiting mass of all but perhaps the stiffest and most rapidly rotating neutron stars. In my opinion, the existence of stellar black holes in binary systems has been confirmed with about 99% certainty; there are no plausible alternative explanations for the observations.

Probable black holes have also been found in the centers of a few specific galaxies, although in most cases the evidence is not yet as convincing as it is for the stellar binary systems. The gas and stars in the nuclei of some galaxies appear to be moving very rapidly, as though a powerful force (almost certainly gravity) were pulling on them. It is conceivable that a central, very tightly packed cluster of stars can provide the requisite mass, but a more plausible alternative is a giant black hole whose mass is 10^6 to 10^9 $M_\odot$. Recent data provided by the HST strongly support the conclusion (made 2 decades ago as a result of ground-based observations) that a supermassive black hole of about 2 billion solar masses is present in the center of the elliptical galaxy M87, 50 million light-years away in the Virgo cluster — gas is moving very rapidly around the center. Even better evidence is found for the spiral galaxy NGC 4258, 20 million light-years away — precise radio measurements reveal a disk of gas in a high-speed circular orbit consistent with the presence of a supermassive (~ 3×10^7 $M_\odot$) black hole. Perhaps the best evidence for a giant black hole was very recently found in the galaxy MCG–6–30–15 with the X-ray satellite ASCA, a collaborative effort between Japan and the U.S. — analysis of a strong iron emission line shows that gas is moving at speeds of up to $0.3c$ in the inner parts of an **accretion** disk.

M87, NGC 4258, and MCG–6–30–15 are **active galaxies** — galaxies whose nuclei emit radiation that cannot be produced by stellar processes alone. In some cases, the so-called **quasars**, the nuclei can be 10 to 1000 times more powerful than the entire galaxy of normal stars. It has long been conjectured that supermassive black holes accreting matter from their surroundings are the central engines of active galaxies. Before being swallowed, the gas can liberate a tremendous amount of energy as it falls toward the black hole — perhaps the equivalent of 10% of its rest-mass energy. This is about 10 times more efficient than the conversion of hydrogen to helium through nuclear fusion in stars. The arguments for the presence of black holes in active galaxies are very persuasive, but not yet absolutely definitive; thus, we eagerly look forward to even higher quality measurements of the motions of stars and gas in galactic nuclei.

Myths about Black Holes

There are some popular myths concerning black holes. The first is that black holes suck up everything in sight, like giant cosmic vacuum cleaners. This is not the case. A black hole's range of influence is limited; only objects in its immediate vicinity will be strongly pulled toward the hole, and even then it is possible to achieve stable (or nearly stable) orbits a safe distance away. For example, if the Sun were to somehow be transformed into a black hole (perhaps by an enormous vise), Earth's orbit would not be altered; the masses of the Sun and Earth would remain constant, as would the distance between them, so the force (according to Newton's law of gravitation) would be

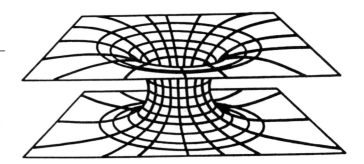

FIGURE 4.19

Schematic diagram of two nonrotating black holes connected by a wormhole. The curved lines denote the curvature of space near a black hole in this two-dimensional spatial "slice" of four-dimensional space-time.

unchanged. Indeed, the gravitational field would remain the same everywhere outside the current radius of the Sun. Only at radii smaller than that of the Sun would the force be stronger.

Second, black holes do not form anywhere and for no apparent reason, as is sometimes implied by cartoons. They might be produced 1) by very massive stars, 2) by neutron stars accreting mass from gravitationally bound companions, 3) in the centers of galaxies, and 4) perhaps by inhomogeneities in the density of matter shortly after the Big Bang. In other situations they are very difficult to make. Our Sun, for example, definitely will not turn into a black hole.

Third, travel to other universes, or to other parts of our Universe, is probably not possible, at least for macroscopic objects. This misconception arises in part from diagrams such as that in Figure 4.19 — one black hole is connected to another black hole by a tunnel, or **wormhole** (officially called an Einstein-Rosen bridge), and it appears possible to go directly through this tunnel. However, the map is misleading; it does not adequately describe the structure of space-time inside a black hole. More detailed analysis shows that to survive a passage through a nonrotating black hole, an object would have to exceed the speed of light, which is impossible.

The geometry of a rotating or a charged black hole is quite different, on the other hand, and travel through one at subluminal speeds initially seems feasible. However, the favorable geometry is only valid for an idealized black hole into which no material is falling; as soon as an object actually tries to traverse the wormhole, the throat closes. One needs to have a very exotic form of matter to keep it open, and there is no evidence for the existence of matter having such properties, at least not in measurable quantities. (The matter must have a negative energy density in the reference frame of a beam of light passing through the wormhole.) Thus, travel via black holes to other universes or other parts of our Universe is probably not possible, despite the allure of such a concept for science-fiction writers.

CONCLUSION

Astronomers are fascinated with supernovae, neutron stars, and black holes. We have learned much during the past few decades, but our understanding of the endpoints of stellar evolution is still far from complete. Stay tuned!

FURTHER READING

Stellar Evolution and White Dwarfs

Cooke, D. A. 1985. *The Life and Death of Stars* (New York: Crown).

Kawaler, S. D., and Winget, D. E. 1987. White dwarfs: Fossil stars. *Sky & Telescope 74(2)*: 132–135.

Kippenhahn, R. 1993. *100 Billion Suns: The Birth, Life, and Death of the Stars* (Princeton: Princeton University Press).

Meadows, A. J. 1978. *Stellar Evolution* (2nd ed.) (Oxford: Pergamon Press).

Sagan, C. 1980. *Cosmos* (New York: Random House).

Trimble, V. 1986. White dwarfs: The once and future suns. *Sky & Telescope 72(4)*: 348–353.

Supernovae

Bethe, H., and Brown, G. 1985. How a supernova explodes. *Scientific American 252(5)*: 60–68.

Evans, R. 1989. Supernova hunter. *Astronomy 17(11)*: 94–97.

Filippenko, A. V. 1993. A supernova with an identity crisis. *Sky & Telescope 86(6)*: 30–36.

Goldsmith, D. 1989. *Supernova!* (New York: St. Martin's Press).

Kirshner, R. P. 1988. Supernova: Death of a star. *National Geographic 173(3)*: 618–647.

Kirshner, R. P. 1994. The Earth's elements. *Scientific American 271(4)*: 58–65.

Marschall, L. A. 1994. *The Supernova Story* (Princeton: Princeton University Press).

Talcott, R. 1988. Insight into star death. *Astronomy 16(2)*: 6–23.

Woosley, S. E., and Phillips, M. M. 1988. Supernova 1987A! *Science 240*: 750–759.

Woosley, S., and Weaver, T. 1989. The great supernova of 1987. *Scientific American 261(2)*: 32–40.

Neutron Stars and Pulsars

Bailyn, C. 1991. Problems with pulsars. *Mercury XX(2)*: 55–60.

Greenstein, G. 1983. *Frozen Star* (New York: Freundlich Books).

Greenstein, G. 1985. Neutron stars and the discovery of pulsars. *Mercury XIV(2)*: 34–39, 62 and *XIV(3)*: 66–73.

Hewish, A. 1989. Pulsars after 20 years. *Mercury XVIII(1)*: 12–15.

Seward, F. 1986. Neutron stars in supernova remnants. *Sky & Telescope 71(1)*: 6–10.

van den Heuvel, P. J., and van Paradijs, J. 1993. X-ray binaries. *Scientific American 269(4)*: 64–70.

Vershuur, G. 1988. On the trail of exotic pulsars. *Astronomy 16(12)*: 22–31.

Will, C. 1989. The binary pulsar: Gravity waves exist. *Mercury XVI(6)*: 162–173.

Black Holes

Hawking, S. W. 1988. *A Brief History of Time* (New York: Bantam Books).

Kaufmann, W. J. III. 1973. *Relativity and Cosmology* (New York: Harper & Row).

Kaufmann, W. J. III. 1979. *Black Holes and Warped Spacetime* (New York: W. H. Freeman).

McClintock, J. 1988. Do black holes exist? *Sky & Telescope 75(1)*: 28–33.

Price, R., and Thorne, K. 1988. The membrane paradigm for black holes. *Scientific American 258(4)*: 69–77.

Shipman, H. L. 1980. *Black Holes, Quasars, and the Universe* (2nd ed.) (Boston: Houghton Mifflin).

Thorne, K. S. 1994. *Black Holes and Time Warps: Einstein's Outrageous Legacy* (New York: W. W. Norton & Co.).

CHAPTER 5

THE ORIGIN AND EVOLUTION OF THE CHEMICAL ELEMENTS

Virginia Trimble

INTRODUCTION

Life on Earth is mostly just very complex chemistry. The **chemical reactions** that go on in your body — and in your dog, your potted philodendron, and the mildew in your shower — turn molecules of food, water, and air into flesh and blood, meanwhile producing the energy that keeps you going. To quote Nobel Prize winning physicist Richard Feynman, "yesterday's mashed potatoes are tomorrow's brains" (and this is more obvious in some people than in others).

Not by chance, the chemical elements most important in living creatures are, with a few exceptions, the commonest ones in the Universe, beginning with hydrogen and oxygen. Their compound, water, makes up half or more of the weight of most cells. Next is carbon, whose unique ability to combine in many different ways with many other atoms makes it the building block of most organic molecules, including carbohydrates and fats. Nitrogen is an essential part of proteins. Other common, important elements are the iron in your blood and the calcium and phosphorus in your bones.

Some less common elements, like manganese, selenium, magnesium, sodium, chlorine, and potassium, are also essential for long-term health and are found in a variety of plant and animal tissues. Sulfur contributes to the coloration of some plants and bacteria (not all of which cause diseases!). The only common chemical elements not found in living cells are helium, neon, and argon, which form no compounds of any kind, and so are called *noble gases*.

Clearly, life would have developed very differently (if at all), if some different set of elements were the common ones on Earth and in the Universe. Thus it becomes interesting to ask why there is lots of oxygen and carbon, but very little beryllium and fluorine. Along the way, we will also find out where gold and silver, uranium and thorium, and all the other elements come from.

Since this is not a detective story, the answer comes first. Hydrogen and helium are left from the very earliest days of the Universe, billions of years ago. Virtually all the other elements are products of **nuclear reactions** in stars, and the supply of them has been building up gradually. We understand quite reasonably well which reactions occur where and when, and what their products are. The process of **nucleosynthesis**, the building up of heavy, complicated atoms from light, simple ones (mostly hydrogen and helium) is, therefore, one of the better-understood parts of modern astrophysics.

The following sections will explore the various astronomical sites of nucleosynthesis and what is produced in each, ending with an overview of the evolution of the chemical composition of our galaxy and the implications for the numbers, ages, and locations of potentially habitable planets. Figure 5.1 shows the measured abundances of the elements, as a function of the number of protons in the nuclei of their atoms, averaged over our galaxy. If we have understood the origin and evolution of the chemical elements, then all the processes of nucleosynthesis will add up to this average. By and large they do.

By way of reminder, chemical reactions merely move atoms of particular elements in and out of molecules of particular compounds (for instance, burning coal or carbon in air or oxygen to make carbon dioxide). Nuclear reactions, in contrast, turn one element into another, as when three helium atoms fuse to make a carbon atom. If you care to look inside the atoms (see Figures 5.2A, 5.2B, and 5.2C) you will see that chemical reactions involve borrowing, lending, and sharing of electrons in the outer parts of atoms, while nuclear reactions attack the protons and neutrons inside, sometimes even turning one into the other.

BIG BANG NUCLEOSYNTHESIS

Ten or twenty billion years ago, the Universe went through a hot, dense phase that largely wiped out any evidence for what might have happened before. This hot, dense phase is called the **Big Bang**, and there is really no escaping the conclusion that it happened (see Chapter 1).

At one time, many scientists thought (Gamow, 1949) that all of the elements, in the proportions we see them, might have come from the Big Bang, which was then pictured as having matter exclusively in the form of **neutrons** (particles with zero charge that are stable only when inside a nucleus). The neutrons (n's) would begin to decay into **protons** (p's) and **electrons** (e's), the positively and negatively charged particles in atoms, which would then combine with remaining neutrons to make the various elements. The picture seemed to work because, especially among the elements heavier than iron, the relative abundances of different elements and isotopes are inversely proportional to how likely each one is to capture a neutron and turn into something else. **Isotopes** are forms of the same element that have different numbers of neutrons in their centers. Carbon, for instance, exists as stable ^{12}C (the common kind) and ^{13}C (about 1% of the carbon in your body) and unstable ^{14}C, with 6, 7, or 8 neutrons, respectively, and, always, 6 protons (which is what makes it carbon). ^{14}C decays to nitrogen-14 in about 6000 years, when one of its neutrons turns into a proton.

This simple, elegant picture, in which all the elements came from the early, hot, dense, Universe, collapsed at the prick of a seemingly tiny pin — the non-existence of nuclei with either 5 or 8 particles. It is easy to take one proton (the same as a nucleus of ordinary hydrogen) and add to it, one particle at a time, to make ^{2}H (deuterium), helium-3 (extraordinarily rare and expensive on Earth), and ^{4}He (the common kind).

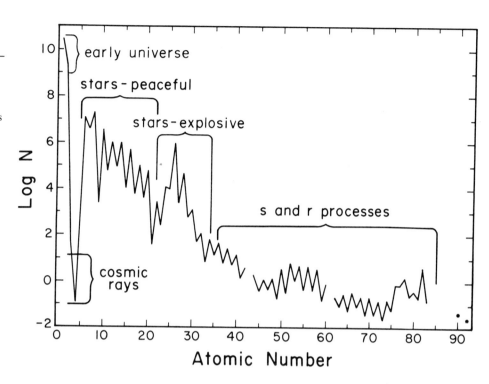

FIGURE 5.1

Relative abundances of the naturally occurring long-lived and stable elements, from hydrogen at the left to uranium at the right. The vertical axis is logarithmic, so that hydrogen is a million-million times as common as beryllium (atomic number = 4). The types and locations of the nuclear reactions that produced each group of elements are shown.

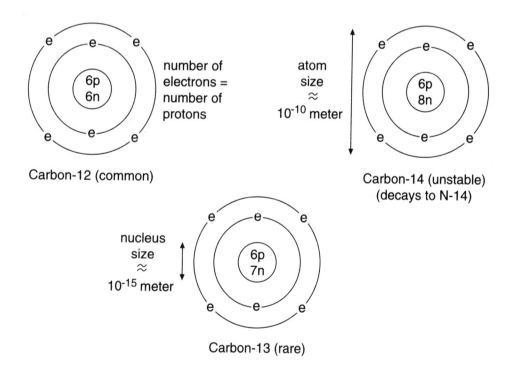

FIGURE 5.2A

Structure of elements, atoms, and isotopes. Each element consists of atoms having a particular number of protons in their nuclei. Carbon has six. Isotopes are variants of elements with different numbers of neutrons in the nuclei of their atoms. Common carbon has six neutrons; a rarer kind has seven; and you can tell organic from inorganic carbon in old rocks by the amount of each. An unstable isotope, carbon-14, decays to nitrogen-14 in about 5700 years. It is made by cosmic rays hitting the Earth's atmosphere and is used to date archaeological sites.

FIGURE 5.2B

Chemical reactions, including biochemical ones, are reactions in which only the electrons are involved. They can be lent, borrowed, or shared, until the outer shell contains a stable number of electrons (2, 8, 8 or 18, depending on the element). The atoms involved are then held together by the unbalanced charges. In this example, an atom of chlorine takes an electron from an atom of sodium, leaving each with a closed outer shell of eight electrons. The plus (+) charge on the sodium and the minus (−) charge on the chlorine hold them together in a molecule of sodium chloride (ordinary salt).

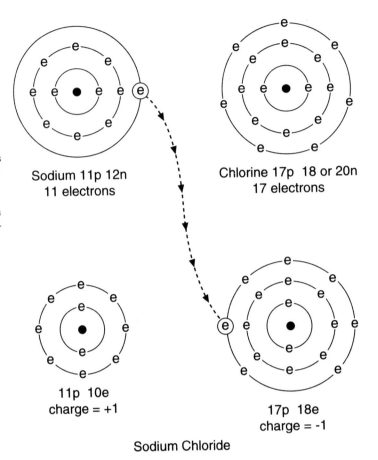

FIGURE 5.2C

Nuclear reactions are ones in which protons and neutrons move between atomic nuclei or even change into each other. In this example, a helium-4 strikes a carbon-13; the particles are thoroughly entangled; and there emerge an oxygen-16 and a single neutron. The neutron is unstable and will decay to a proton + electron + neutrino unless it is captured by another nucleus within about 11 minutes. This reaction is the source of neutrons for a subset of the capture by iron (etc.) that build up the heaviest elements.

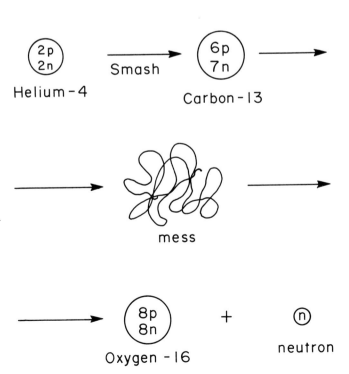

But then what? You cannot add another n or p to helium-4 and make anything stable, nor can two helium-4 nuclei fuse. The process stops, having made "all the elements up to helium" and nothing else.

Actually, I am lying to you. A very rare pair of reactions makes a tiny amount of lithium-7. If the early Universe had been much denser when it was at the right temperature for nuclear reactions (near 10 billion K) then, indeed, a much wider range of reactions could have taken place, making many more elements. The elements would not, however, come out in the proportions we see. In addition, such a universe would have a very short lifetime and be uninhabitable for other reasons.

Big Bang nucleosynthesis is, nevertheless, important for two reasons. First, the amounts of things produced agree quite well with the abundances we see in gas and stars that have not been much subjected to later processes — about 3/4 hydrogen and 1/4 helium with one or two atoms of ^{2}H and ^{3}He in every hundred thousand, and one Li among 10 billion atoms. This adds to our confidence that the basic Big Bang picture is correct.

Second, the precise amounts of deuterium, helium-3 and -4, and lithium-7 left by the hot, dense stage is a very sensitive probe of just how hot and dense it was and of how fast the Universe went through that stage. As a result, we can say fairly firmly that (a) the total amount of ordinary matter in the Universe, including all the elements, is much smaller than the amount needed for its gravitational attraction to win out over and stop the present expansion of the Universe (see Figure 5.3) and (b) no additional kinds of particles beyond the ones we already know in the laboratory could have existed during the nucleosynthesis phase.

Chapter 1 explains these results in greater detail. Both of them are important for understanding the origin of galaxies (see Chapter 2). A very large fraction of the material whose gravitational forces pull together and assemble the galaxies out of diffuse material is dark. Because it neither emits nor absorbs light, we cannot learn about its composition by analyzing the light from it, the way we do with ordinary stars and gas. But thinking about Big Bang nucleosynthesis does tell us that the **dark matter** is not mostly ordinary stuff, and it is not certain kinds of hypothetical, exotic particles.

ELEMENT SYNTHESIS IN STARS

To progress beyond hydrogen and helium, we must find some other location that is about as hot and dense as the early Universe. One such place is a nuclear reactor or accelerator in a physics laboratory. And, indeed, virtually all of the reactions discussed here can be duplicated in laboratory experiments. The results of the experiments are measurements of how fast particular reactions go at different temperatures and densities, the amount of energy released (or absorbed) in the reactions, and the particular elements made by each. These results fit nicely into a theoretical framework that comes from a general understanding of the forces that hold protons and neutrons together in nuclei. And the product elements are the ones we see in stars. This sort of interplay among experimental results, theory, and observations is characteristic of the way astronomy and other sciences progress. Because things fit together, we can be confident we understand which nuclear reactions will occur under any set of conditions and what they will produce.

A third place that is hot and dense enough for nuclear reactions is the center of a star. That stars derive most of their energy for most of their lives from nuclear reactions

FIGURE 5.3

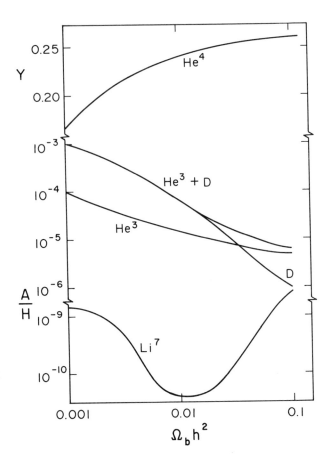

Abundances of the elements and isotopes produced in the hot, dense early Universe (Big Bang) as a function of the density of ordinary matter participating in the reactions. On the horizontal axis, 0.01 means 1% of the density needed to stop the expansion of the Universe, for a Hubble constant of 100 km/s/Mpc (h = 1). Helium-4 is shown as a fraction, Y, of the total mass of ordinary matter (a little less than 25%), and the others as ratios of numbers of their atoms (A) to numbers of ordinary hydrogen (H), for deuterium (hydrogen-2, also called D), helium-3, and lithium-7. The lithium curve has its odd shape because the element is made by two different reactions under different conditions. Observed values of the abundances of ^{2}H, ^{3}He, ^{4}He, and ^{7}Li in material that has not been exposed to nuclear reactions in stars show that the density of ordinary matter (now hydrogen, helium, carbon, and all the rest) is much less than would be needed to stop the expansion of the Universe in the distant future.

has been known for about 60 years, and Hans Bethe (Nobel Prize in physics, 1967) wrote down the details of the most important ones in the late 1930s (Bethe and Critchfield, 1938; Bethe, 1939) (Figure 5.4).

Understanding of the more complicated reactions that make the heaviest elements came more slowly after World War II. A key event was the 1957 publication, by Burbidge et al. (1957), of a massive review of observed element abundances in stars and of the full range of nuclear reactions needed to make them all. This paper is important enough in astronomers' minds that it is instantly recognized by the acronym B^{2}FH. Since then, studies of nucleosynthesis have fine-tuned their results, and more recent reviews of the subject (Trimble, 1975, 1991) normally build upon the framework erected by B^{2}FH. Even now, more precise numbers are still needed for a few reactions (for instance carbon + helium making oxygen) and are being actively pursued by experimenters and theorists in many countries.

The Major Burning Phases—Hydrogen

Let's start with an outline (see Table 5.1). It will be easy to remember if you keep in mind that the lightest nuclei, with the fewest particles in them, generally fuse at the lowest temperatures (beyond iron you cannot fuse them at all), and stars evolve with time in the direction of having hotter, denser cores.

FIGURE 5.4

Stars can live for billions of years on hydrogen fusion precisely because the Big Bang left 75% or more of the matter in the form of hydrogen. That ordinary stars are made mostly of hydrogen and helium was first shown in the Ph.D. dissertation of the woman shown here, Cecilia Payne, later Payne-Gaposchkin. It is exceedingly rare for a scientist to make a fundamental discovery as part of a thesis project (though you will hear about one other case later), and Payne's results were not fully believed by the community until the dominance of hydrogen had been demonstrated by another, more senior astronomer. The picture, which resembles a Coptic tomb portrait, was drawn by her husband, Sergei Gaposchkin, in about 1945. (Courtesy of Katherine Gaposchkin Haramundanis.)

The first set of reactions, and the ones that keep stars shining for 90% or more of their lives, simply fuse hydrogen into additional helium. Next, helium burns (a standard synonym for **fusion** reactions) into carbon and oxygen. Our Sun will progress no further. In more massive stars, carbon- and oxygen-burning makes neon, magnesium, silicon, and other intermediate mass elements, which, in turn, fuse to iron and its neighbors (nickel, copper, and others). At this point, the star is in serious trouble. So far, each reaction stage has released energy to keep the star hot and shining. The protons and neutrons in iron nuclei are grouped as tightly together as the forces permit, and no fur-

TABLE 5.1 Stellar Evolutionary Stages

Reactions	Products	Temperature of burning (K)	Time scale for Sun (yr)	Time scale for star of 20 times mass of Sun (yr)
Hydrogen burning	Helium	$1-4 \times 10^7$	10^{10}	10^7
Helium burning	C, O	$1-2 \times 10^8$	10^9	10^6
Carbon burning	Ne, Na, Mg	8×10^8	—	300
Neon burning	Mg, Si	1.7×10^9	—	< 1
Oxygen burning	Si, S	2.1×10^9	—	< 1
Silicon burning	Ti to Zn	4×10^9	—	2 days

ther nuclear energy can be extracted. The outcome is one kind of supernova (see Chapter 4). And the building of elements beyond iron occurs in a rather different way.

Hydrogen burns in two separate reaction sequences, both known to Bethe. In the simpler, two protons come together, one turns into a neutron, and they stick together as a deuterium. The deuterium nucleus quickly picks up another proton, and, after a few more particle collisions, you have a helium-4 nucleus, and lots of energy, which the star then radiates. This **proton-proton** (or p-p) **chain** has been the main energy source in our Sun for the past 4.5 billion years, and will continue to be so for the next 5 billion or thereabouts. Two interesting sidelights: First, the p-p chain begins by turning a proton into a neutron. This is a very slow reaction, and the Sun therefore burns its hydrogen quiescently. In hydrogen bombs, you start with deuterium, and the rest goes very quickly. Second, turning protons into neutrons releases tiny, uncharged particles called **neutrinos**. These stream directly out from the center of the Sun (while the light takes 100,000 years to get out) and are detected here on Earth. Thus, we know that the Sun is fusing hydrogen right now.

Hydrogen can also burn in what is called the **CNO tri-cycle**, in which atoms of carbon, nitrogen, and oxygen act as catalysts. Of course this can happen only if C, N, and O atoms are already present, which means that the very first generation of stars had to begin with the proton-proton chain in all cases. But stars today have about 1% of their masses in CNO, and the CNO cycle is the main hydrogen burner for stars more than about twice the mass of our Sun. It is interesting because a byproduct is the conversion of some carbon and oxygen into nitrogen, which otherwise does not turn up in our outline, and is vital for **proteins** and **DNA**.

Hydrogen-burning continues until the central 10% of the star is all helium. This takes billions of years for smallish stars like the Sun, but only millions of years for more massive ones. At the same time nuclear reactions are changing the composition of a star's center, its outside reacts by expanding or contracting and changing temperature. We can calculate what those changes will be, and compare them with the brightnesses, sizes, and colors (temperatures) of stars we see at different ages and masses, thereby checking that we have the right answer. Very broadly, an evolving star gets brighter and redder for most of its life, with the possibility of one or more switch-backs to being fainter and/or bluer. Hydrogen-burning stars are said to be on the main sequence, and more evolved ones are called **red giants** and **supergiants**.

The Major Burning Phases — Helium

So far, we have made helium. This was the point at which the Big Bang hung up. What is the difference? The Universe cools as it expands, and particles get further and further apart, so it is harder and harder for them to interact. Stellar cores, in contrast, get hotter and denser. And by the time 10% of the interior has burned to helium, it becomes possible for three helium nuclei all to meet up within a tiny fraction of a second and fuse to form carbon.

Helium-burning starts explosively in low-mass stars and shakes them up, but starts peacefully in massive stars. Similar cases occur later, so we might as well bite the bullet and understand why right now. Ordinary hot gases try to push outward and expand; and the hotter they are, the harder they push. Normal nuclear reactions release energy and heat the burning gas. If a reaction goes too fast, the gas overheats and expands, cooling back down again to stability. But in very dense gases, the amount of pushing does not depend on temperature. Such gases are said to be **degenerate** (a statement about how their electrons are moving, not about their moral principles). If you turn on a nuclear reaction in a degenerate gas, the gas heats up, but does not expand. The hotter gas burns faster, and so forth, and pretty soon you have a nasty explosion on your hands (or at least in your star). It is the low-mass stars that start helium-burning with a flash, because they are denser than big stars. This probably sounds backwards, but it really is the right answer.

Helium-burning has two parts that go on together. Three He nuclei fuse to a single ^{12}C; and ^{12}C plus ^{4}He make oxygen-16. The rate of this second part is very important (and not very well known), because the relative amounts of C and O made determine what is available to burn later. That you get some of each is vital to our existence on earth — we need both carbon-based food and oxygen-rich air!

At the same time helium-burning is making C and O for us in stellar cores, hydrogen fusion continues further out. The stars are by now very bright and are blowing large amounts of gas off their surfaces in winds we can see (even the Sun has a very weak one). Both nuclear reactions and the vigorous loss of mass go on until one of two things happens. Either the C-O core gets hot enough for further reactions (of which more in a moment), or wind loss strips the star down to its core, which then begins to cool, turning off hydrogen- and helium-burning. Additional reactions are the fate of stars that begin life with more than about 8 times the mass of our Sun, and stripping ends the lives of the others.

Color Plate 4.1 shows a product of the stripping operation. The hot, remnant central core illuminates surrounding material blown out in the wind. The gas cloud is called a **planetary nebula** for misleading historical reasons (nothing to do with whether the star had planets). The core will cool to what is called a **white dwarf**, and, in general, do nothing else interesting for the rest of the history of the Universe.

From a nucleosynthetic viewpoint, most low-mass stars are rather a disappointment. They keep most of the C and O they have made inside the white dwarf. The wind does, however, carry off some carbon and nitrogen, and perhaps oxygen, and products of one minor set of reactions (see below) and dump them back into the general interstellar medium, where new stars are forming (see Chapter 3).

Very rarely, a white dwarf (or a pair of them) left behind by dying, low-mass stars, gets into trouble. No white dwarf can exceed a mass 1.4 times that of our Sun, or it will either collapse or start nuclear burning and explode (because it is degenerate; I told you we would need this again!). A previously stable white dwarf can find itself above the limit if another star dumps material onto it or if a close pair of white dwarfs merge. In either case, the result will be a kind of supernova (see Chapter 4). The explosive burning turns most of the white dwarf into iron and related elements. Such events may be the main source of iron (plus chromium, cobalt, etc.) at some times and places in the Universe.

The Major Burning Phases — Heavy Elements

In very massive stars, wind stripping still occurs. But it does not prevent the onset of carbon burning. From now on, the star uses up its fuels so quickly that the outer layers never find out what is going on inside. The star continues to look like a red or blue supergiant, whichever it was when carbon fusion started. At successively higher temperatures and with successively shorter time scales, carbon burning is followed by neon-, oxygen-, and silicon-burning. The durations are brief, partly because these reactions release rather little energy compared to hydrogen- and helium-burning, and partly because the very hot stellar cores radiate enormous numbers of neutrinos, which drain away energy without contributing to starlight.

Another factor common to the reactions that burn heavy elements is that all are complex networks rather than simple chains or cycles. Several different elements, as well as free protons, neutrons, and helium nuclei, are all present at once and can interact in many different ways. Each burning stage thus makes many different elements and isotopes, of which only the commonest are mentioned in Table 5.1. As each new reaction turns on at the center of the star, the others move outward into slightly cooler regions and continue to operate, until the star looks rather like an onion (see Color Plate 4.11).

The onion skins work their way outward through the star until the mass of the central iron (etc.) core grows to a little more than the mass of our Sun (remember this is a star that began with 10 or 20 or 30 solar masses). Catastrophe is at hand. There is a maximum stable mass for the degenerate iron core, just as there is for degenerate white dwarfs. The limit is called the **Chandrasekhar mass**, curiously enough because it was discovered by Chandrasekhar (1931, 1935) (You can't count on these things — Hubble's law, Chapter 1, was discovered by K. Lundmark.) In any case, Chandrasekhar shares some tiny part of the blame for this article, because he was the teacher of the present author's thesis advisor. Chandra's Nobel Prize arrived 50-some years after his landmark paper was published.

A vital difference between massive stars and the ones that make white dwarfs is that the white dwarf keeps most of the carbon and oxygen made by its parent star, while supernova explosions of more massive stars blow off most of the heavy elements they have made. The precise amounts of each element and isotope in the onion, waiting to be blown out, depend on the mass of the star and other factors. In addition, those amounts will change after the shock of the explosion passes through the layers, generally making a bit more of the less common isotopes. But the general patterns are clear and agree with what we see as the composition of the Sun and stars.

On average, you get the most of the things that form first. Next, elements with even numbers of protons are commoner than ones with odd numbers. Neon outnumbers fluorine and sodium; silicon outnumbers aluminum and phosphorus. Isotopes with even numbers of both protons and neutrons (ones you could think of as being put together from a bunch of helium nuclei) are particularly abundant — ^{16}O, ^{20}Ne, ^{24}Mg, and ^{28}Si. And there is lots of ^{56}Fe. These patterns make sense. The commonest species are the ones that, according to our understanding of nuclear forces, have their protons and neutrons held together most tightly. Similar patterns will appear among the heavier elements discussed in the next section.

Massive stars are much more important for nucleosynthesis than smaller ones like the Sun for three reasons. First, they experience a much wider variety of nuclear reactions with a much wider range of products. Second, at the ends of their lives they blow out most of the heavy elements they have made in supernova explosions, whereas low-mass stars keep most of their products in the C-O white dwarf remnant. Third, their lives are much shorter. Thus many generations of 30-solar-mass stars lived and died before our solar system ever formed, and many more have formed and exploded since (see Figure 5.5).

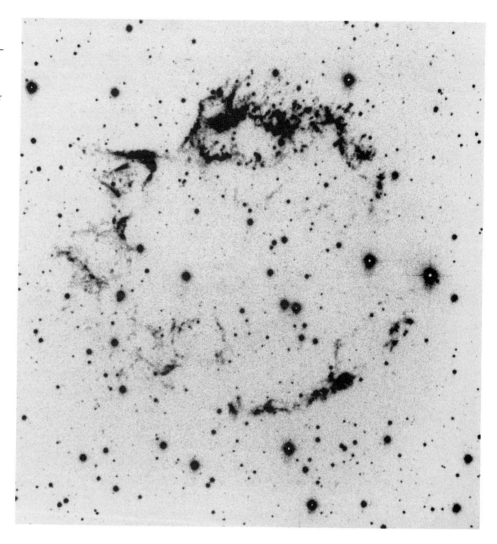

FIGURE 5.5

Remnant of a recent supernova. This expanding gas cloud, Cassiopeia A, is the debris from a very massive star that exploded a little more than 300 years ago (the supernova event itself was not definitely seen). Different gas blobs are rich in different elements, including nitrogen, oxygen, and argon. This photograph highlights emission by sulfur. (Courtesy of Robert A. Fesen.)

On Beyond Iron

In a sense, we are almost through. Elements whose origins you have now heard about make up 99.99997% of all the matter in the Sun and other stars. On the other hand, we have made only the first 30 elements, and there are another 62 to go, before we get to uranium. Admittedly, not all are stable, and so not all are currently found in the Sun or Earth or meteorites. On beyond uranium are 18 or more short-lived elements that have been produced in terrestrial laboratories. Of these, at least plutonium was present when the Earth first formed and so must also have been made in nature.

Where did these few, but highly varied, atoms come from? Ordinary fusion reactions are no longer relevant. Throw two iron nuclei (each with 26 protons) at each other, and they will either bounce off or break each other up, not fuse to make tellurium, the element with 52 protons. As B^{2}FH showed, three additional processes are needed to make the full range of elements beyond number 30. In two of them, seed nuclei, mostly iron, capture neutrons, painfully, one by one, with neutrons occasionally decaying to protons, to keep the nuclei stable. A third process acts on the products of these neutron captures (from germanium to lead) and either adds protons or knocks away a few neutrons. No wonder these products are rare! Rarest of all are those from the third process, ranging down to tantalum-180, of which there is one atom for every 10^{16} atoms of hydrogen in the Universe!

Recognizing the need for two separate neutron capture processes was a real triumph, also achieved by Cameron (1957), though his version was not publicly available until later than B²FH. Remember Gamow, who wanted to make everything out of neutrons? His evidence for this was the correlation between abundances of elements and isotopes and their willingness to capture another neutron. Unwilling capturers pile up and become abundant, while willing ones are quickly turned into other species and remain rare.

But careful examination of abundances of elements 30 to 92 reveals two patterns. Some relatively common isotopes are unwilling to capture neutrons in their present condition. They have 50, 82, or 126 neutrons now ("magic" numbers because they are closed neutron shells, analogous to shells of 8 or 18 electrons). Barium-138 and lead-208 are examples. Other relatively common isotopes would happily capture neutrons now, if they had a chance. But suppose they formed by grabbing every neutron in sight, while they could, long ago, until they had 50 or 82 or 126, and then some of those neutrons later decayed to protons, leaving atoms of greater long-term stability. This would account for the relatively large amounts of tin and tellurium, rhenium, osmium, and platinum now found. I keep emphasizing the "relatively" because even the commonest of these elements makes up less than one atom in a hundred million.

B²FH called the two necessary processes s (for slow neutron capture) and r (for rapid neutron capture), where slow and rapid mean in comparison to the time it takes an unstable nucleus to decay back to a stable one, by having a neutron turn into a proton. This time ranges from minutes to years. The third, fine tuning, process is generally called p, for proton. How and when and where do these three happen?

The sites of the r- and p-processes are apparently various kinds of supernova explosions, when indeed both lots of iron and lots of neutrons will be available together, at least for a short time. Just how you get the products out of the explosions without damaging them is a topic of current debate. Uranium and thorium, without which atomic bombs could never have been developed, are among the elements made only by the r-process (oh, sorry; I said I wouldn't mention bombs again!).

The s-process is, in some ways, more interesting. It happens toward the end of helium-burning in red giants and supergiants of masses from 1 to 10 or so solar masses, including, that is, our own Sun, 5 billion years from now. The trick is that the star mixes some ^{14}N from the hydrogen-burning zone into the helium-burning zone, where the ^{14}N is converted to neon-22. One more helium nucleus zaps the ^{22}Ne and knocks off a neutron. Carbon-13 (also made in hydrogen-burning) will act the same way. The liberated neutrons wander around until they meet some fairly heavy nucleus, which then captures them. And, voila, s-process! Of course this works (like the CNO cycle) only in stars that already had some heavy elements when they formed. In contrast, the r-process works on iron made a little earlier in the same (massive) star and so could begin in the very first generation of stars.

Thus, the third contribution of low-mass stars to nucleosynthesis, besides carbon, nitrogen, and oxygen (and iron when white dwarfs explode) is the shedding of s-process products in their winds and planetary nebulae. One such product, technetium, has no stable isotopes and lives only a million years. It was the discovery of Tc in a few highly evolved stars by Merrill (1952) that showed conclusively that complex nuclear reactions are going on right in front of our eyes (or telescopes).

A Few Loose Ends

So far, so good. The Big Bang made hydrogen and helium. Elements from carbon to zinc arise in peaceful nuclear reactions, mostly in massive stars. And the heaviest ones come from the s-, r-, and p-processes acting on iron, cobalt, nickel, and so forth that are already present.

Are we home clear? Almost, apart from that lower left corner of Figure 5.1. The Big Bang left behind a little bit of lithium-7 (perhaps 1% of what we see in young stars). But nowhere in the preceding paragraphs have we yet found processes that make any beryllium or boron, or the rest of the lithium. There is a good reason for this. All three are very fragile. Inside a star, they will quickly burn to helium and other sturdier elements. Like deuterium, they are destroyed in stars, not made there (except maybe lithium, which is strangely common in a very few red giants).

B^2FH attributed lithium, beryllium, and boron to an unknown x-process. We are now reasonably sure that they are by-products of interstellar traffic accidents. The space between the stars is not empty. Rather, it is filled with diffuse interstellar gas (see Chapter 3). In addition, **cosmic rays** fill the galaxy. These are particles, mostly protons, but also nuclei of all the other elements, moving very close to the speed of light and so carrying large amounts of energy (up to as much as a well-thrown baseball, in a single particle). Cosmic rays ultimately derive their energy from supernovae, though some of the details are fuzzy (see Chapter 4).

From time to time, a cosmic ray hits an interstellar atom and busts it up. The process is called **spallation**, and when the victim is a carbon or oxygen nucleus, lithium, beryllium, and boron are among the products. We know this happens because the cosmic rays themselves, when they get to us, contain large excesses of Li, Be, and B, made by the converse process of a cosmic ray C or O being zapped by an interstellar proton and losing portions of its anatomy in the collision. Some other kinds of atoms, including fluorine and sodium, are commoner in cosmic rays than elsewhere and must also be partly made by spallation.

You may never have thought much about these light elements, but lithium is used in treatment of some illnesses, beryllium is a light metal employed in aircraft, and one boron compound, borax, is a traditional cleaning product.

We have now examined every element and isotope, at least briefly, and looked at most of the sites where nuclear reactions ought to occur. Nearly everything seems to be coming out even. One further site must be relevant. Minor explosions — ones that don't destroy the star — occur in some stellar pairs, when a normal (**main-sequence**) star drizzles hydrogen from its atmosphere onto a white dwarf companion. The hydrogen builds up for a hundred thousand years or so, and then (because it is degenerate) blows up in hours. The observed phenomenon is called a **nova**, and the dominant reaction is a hot, fast CNO tri-cycle. Isotopes more common in nova ejecta than in normal hydrogen-burning should include carbon-13, nitrogen-15, oxygen-17, and neon-21 (see Figure 5.6).

Another possible nova product is aluminum-26. You cannot go to the drug store and buy a box of ^{26}Al. It decays to magnesium-26 in 720,000 years, so the solar system supply is long since gone. But we know it is present in the interstellar medium (because we see the radiation emitted during the decay — another proof that nucleosynthesis is an everyday affair). And, second, we know it was present when the asteroids and meteorites formed, because we find the product, ^{26}Mg, in strange places. ^{26}Al decay was, therefore, a source of heat in young solar system objects, and it is generally blamed for

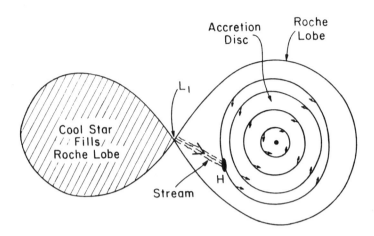

FIGURE 5.6

Anatomy of a nova. Material from a cool star flows in a stream that hits a disk (at H) orbiting a white dwarf (the small black dot at the center of the nearly circular accretion disk pattern). Gas in turn flows from the disk onto the white dwarf until enough has built up so that the bottom of the hydrogen-rich accreted layer is degenerate. The accreted material burns explosively and is blown back off the white dwarf and into space. Some rare isotopes, including perhaps ^{26}Al, are produced primarily in novae. The name *cataclysmic variables*, for novae and related binaries, was coined by Cecilia and Sergei Gaposchkin. Gas will always flow to the second star when the first one is larger than a certain critical volume called the *Roche lobe*. The gas flows most readily through the first Lagrangian point, L1. The lobe and point are named for the French mathematicians who long ago studied the gravitational forces of pairs of stars or other masses.

having melted the asteroids that later broke up to make meteorites. This is why astronomers are interested in where it was made. Novae, supernovae, and the same stars that make s-process elements are all possibilities. Each has both strong proponents and vigorous opponents in the community.

GALACTIC CHEMICAL EVOLUTION

The last two sections described the origin of the individual elements in some detail. In contrast, our picture of the gradual increase in abundances of those elements, from the Big Bang to the time the solar system formed, to the present, must be painted with a much broader brush. Two problems contribute to smearing our image of galactic **chemical evolution**. One is pretty obvious and also has a fairly satisfactory solution. The other is less obvious, and we deal with it by cheating. Among the reasons you might want to solve these problems is that the number of stars with enough heavy elements to have Earth-like planets depends on the progress of chemical evolution.

The first difficulty is that the Milky Way contains about 200 billion stars. As you might guess, the largest computers in the world, even the largest you can imagine, will not keep track of the individual behaviors of 200 billion anythings. For decades, this was thought to be a barrier to serious study of galactic evolution. The problem was solved by a graduate student, Beatrice M. Tinsley, as part of her doctoral research in the 1960s. She divided up stars into about a dozen groups by mass and pretended that all the stars in a given group would do exactly the same thing. Since this is almost true, the models she calculated (Tinsley 1968, 1980) look almost like real galaxies. She had, of course, also to deal with the second problem, and her choice of ways of cheating is often still used.

The second problem comes about because we know less about what a gas cloud wants to do than the cloud itself knows. The basic processes of galactic evolution are easy to list. Interstellar gas and dust turn into stars (see Chapter 3). Nuclear reactions in stars build heavy elements and some of the products are blown back out to mix with remaining interstellar stuff. Fresh gas can flow into the galaxy or winds can expel enriched gas. And gas of different compositions can flow from one place to another inside a galaxy.

Each process is governed by ordinary laws of physics and has a unique outcome. That is, for instance, a particular gas cloud will make a definite number of stars of particular masses at some time. But, even if you tell a theoretical astrophysicist all about the cloud — its distribution of density, temperature, magnetic field, and so forth — he will not be able to tell you in return just when it will form stars, or how many of each mass, or what percentage of them will be in binary pairs. The process is just too messy to calculate.

Similar situations occur in other sciences. Consider weather forecasting. No unknown physics is involved. But, even with a perfect picture of where the clouds are now, which way the wind is blowing in each place, how the ocean currents flow, and so on, you don't plan an outdoor party two weeks in advance on the basis of weather reports. The inherent uncertainty of the situation is reflected in the forecasts themselves: "20% probability of rain," "sunny periods," "rain possibly turning to sleet."

The incalculable items we need for galactic chemical evolution are (a) the total rate of star formation — in solar masses per year, as a function of time, place, and gas composition in the galaxy; (b) the relative numbers of stars of each mass formed (which will also vary with time, place, and composition); (c) rates of gas flow into and out of whatever region you are considering, plus the composition of the gas involved; and (d) how fast enriched material gets mixed with its surroundings.

The standard way of cheating is called an **adjustable parameter**. Consider the rate of star formation. Choose a likely value, for instance the current one in our galaxy (a few solar masses per year). Use it in your model and see if you like the answer. If you don't, vary the parameter called "star formation rate" until you do like the answer. And, if the rate you end up liking is not too unreasonable, you have probably learned something. Parameters representing gas inflow and outflow, the stellar mass distribution, and the rest are treated similarly, one at a time.

Simple Models and the G Dwarf Problem

The model adopted by Tinsley (1968) and many later workers had just about the simplest set of parameters you can think of. They are generally phrased as assumptions or approximations. One is called "instantaneous recycling." This means you pretend that nuclear reactions and mixing of the expelled products occur as soon as the stars form. At least for massive stars, with short lives, this is not unreasonable. The second is called the "one zone" or "homogeneous" approximation. It means that the heavy elements produced by stars get mixed uniformly through all the gas in your models and that star formation comes from gas with this well-mixed composition.

A third approximation is called "constant **initial mass function**." It is equivalent to saying that every gas cloud makes the same assortment of high-, low-, and intermediate-mass stars. You describe the assortment as a power law, with number of stars scaling like M^{-x}, where M is stellar mass and x is a number between 1 and 3 or so. The mix of star masses we see around us now is described at least crudely by x = 2.3. The mix is called the initial mass function or IMF. Thus the simplest assumption is "constant IMF." Important results, like the ratio of carbon to oxygen in your model galaxy, depend on the IMF. The simplest assumption about gas inflows and outflows is that there aren't any. This is the "closed box" approximation.

Finally, star-formation rate is the one parameter that is allowed to vary. The rate could be constant with time, a sharp spike 15 billion years ago with nothing since, or lots of other combinations (see Figure 5.7).

A few years after Tinsley's work, others realized that some of the results of the "homogeneous, instantaneous recycling, closed box, constant IMF" model do not require a computer. You can write down a set of equations and solve them algebraically to learn, for instance, just what percentage of the interstellar gas should be heavy elements when 90% of the gas has been turned into stars. The answer is about 2% heavy elements, which is the correct combination to describe the disk of the Milky Way.

All in all, the biggest surprise connected with the simple models is just how good a job they do. One can reproduce most of the range of galaxy colors, luminosities, compositions, and residual gas contents. The parameter that is allowed to vary, star-formation rate versus time, must be an important one! An initial spike of star formation yields models that resemble elliptical galaxies. Constant star-formation rate makes irregular galaxies like the Large and Small Magellanic Clouds, and a gradually decreasing rate makes spirals like our Milky Way. Relaxing the "closed box" assumption to allow enriched gas to be expelled from low-mass galaxies and low-density regions of bigger ones leads to even better agreement between the models and observations.

Only one conspicuous disagreement between models and data turns up. But it is a fatal one. For galaxies outside our own, we can generally measure only the average chemical composition of the gas or stars in a particular region. Near the Sun, however, we can look at individual stars and count how many have each possible fraction of heavy elements, from 2% to 3% on down to 0.0002%, the smallest found.

The models say that we should find sizable numbers of stars with much less than the solar abundance of heavy elements, provided that we look at a class of stars that live as long as the age of the galactic disk. G dwarfs (including the Sun) live about 10 billion years and so are useful for this study. The "dwarf" part means that they are still burning hydrogen in their cores, and the "G" part means that they have surface temperatures of 5000 to 6000 K (as do all dwarfs with masses close to that of the Sun). G dwarfs with less than 0.2% heavy elements are, in fact, very rare in our neighborhood, much rarer than the models say they should be.

Other less serious disagreements with the simplest models turn up. For instance, the heavy elements do not vary in lock step — stars with small total amounts have higher ratios of oxygen to iron than does the Sun, and so forth. But it was the **G dwarf problem** that drove most attempts at building more complicated models of galactic chemical evolution.

Models with Bells and Whistles

What happens if you begin to violate the assumptions described in the previous subsection? We will get on faster with the discussion if I am allowed one more funny new word. The elements that show up most conspicuously in stellar atmospheres are iron, titanium, sodium, and calcium. These are all metals in the chemical sense. Thus,

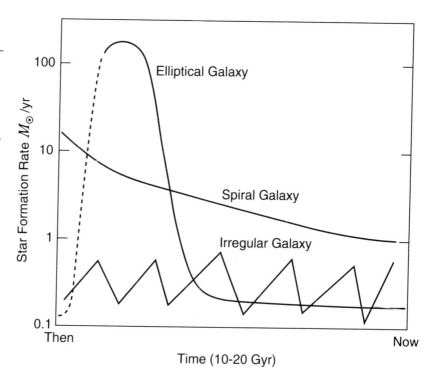

FIGURE 5.7

Star-formation rate in solar masses per year as a function of time in different kinds of galaxies. Elliptical galaxies formed most of their stars very long ago (and may have few or no stars left that can still support life-bearing planets). Spiral galaxies like our Milky Way have formed their stars gradually over time, though not so vigorously now as in the past. Irregular and other dwarf galaxies form their stars in sporadic bursts (probably not so regular as the pattern shown), but still have very low metal abundances.

astronomers decades ago got into the habit of calling all the heavy elements metals and speaking of "metal abundance" and **metallicity** when they meant everything except hydrogen and helium. Of course some of the commonest "metals" are oxygen, neon, and carbon. If you know some chemistry, this will bother you for a while, but I promise you will get used to it.

The one standard approximation that *must* be wrong is instantaneous recycling. Even massive stars live millions of years, and the white dwarf pairs that make some of the iron can live for billions. An honest treatment of the time scales cures many of our minor problems. Old stars, though poor in all heavy elements, have more oxygen than iron because oxygen production started earlier. Non-solar ratios of s-process to r-process elements and of nitrogen and carbon to oxygen in old stars also come out right with proper allowance for stellar lifetimes. We haven't really learned anything new here, except that some kinds of cheating are too blatant to get away with.

The other assumptions and approximations could, however, be true. The G dwarf problem says that they are not all true. The question is then which one or ones to abandon. Curiously, almost any likely variation of the standard model helps. Either outflow of metal-rich gas or inflow of pure hydrogen and helium gas will help to keep the heavy element abundance in the interstellar medium constant through time, so that most G dwarfs will have about the same composition as the Sun. Flows can be either in and out of the galaxy or just in and out of our neighborhood. Observations of gas clouds moving at high speeds indicate that both happen. We don't live in a closed box.

Another set of possibilities comes from looking at details of star formation. It is easier for gas to cool and form stars if it contains largish amounts of metals. High metallicity helps especially in forming low-mass stars. This opens up two alternatives. First, when the galaxy was young and enrichment just starting, stars might have formed mostly in clouds that were very close to supernovae and so had more metals than average. This is metal-enhanced star formation (MESF, pronounced as if it were the capital of some small, middle-eastern country). MESF means that we will see very few stars of low metallicity. It is equivalent to relaxing the "homogeneous" assumption.

Alternatively, or in addition, the earliest stellar generations might have consisted primarily of massive stars, which have long since died. Then most of the G dwarfs we see will have come from gas already enriched in metals by these early, massive stars, solving the problem. This solution is called *variable IMF*, because we are allowing x in the distribution M^{-x} to be different in the past.

To get at the truth, we will eventually have to abandon the adjustable parameter approach to galactic chemical evolution and learn to calculate or predict which gas clouds will form which stars and when, which way the interstellar and intergalactic winds will blow, and so forth. The field appears to need a fresh, new idea, comparable with Tinsley's scheme of dividing up stars into classes by mass, that will permit relatively simple calculations to come closer to representing real galaxies.

You may well be puzzled that the notorious dark matter, responsible for 90% or more of the mass of the Milky Way and other galaxies, has not appeared in this section. As you saw in Chapter 2, dark matter controls how galaxies form. But it genuinely does not get involved in the chemical evolution of gas and stars. There is a good reason for the non-involvement. If the dark matter is very low-mass, faint stars, nothing can yet have happened, because any star with a mass less than half that of the Sun has a life span longer than the age of the Universe. If the dark matter is any of the more exotic candidates, they are not capable of nuclear reactions (or of emitting and absorbing light), so they also cannot get mixed up in chemical evolution.

Stellar Populations and the Incidence of Habitable Planets

Nobody really knows just what is needed for life (even narrowly defined as carbon-based life) to develop. But I'm willing to bet that reasonable amounts of carbon, nitrogen, oxygen, etc. and lots of time are vital. Water, or some other liquid within which chemical reactions can take place, would also be nice. Thus, inhabited, or inhabitable, planets are most likely to orbit stars that are at least a few billion years old and have, at least, say, one-third of the solar allotment of heavy elements. Only stars less than 1.5 times the mass of the Sun can live this long. And only ones more massive than about 0.7 solar masses are bright enough to keep their planets above freezing. Where shall we look for such stars?

Models of galactic chemical evolution, in their current, adjustable parameters state of incompleteness, can manufacture almost any combination of stellar ages, masses, compositions, and locations you ask for. Thus, the question of how many stars in our galaxy are capable of hosting Earth-like planets must be asked observationally. The answer comes out in terms of a concept of stellar populations.

The ages we measure for stars, their metallicities, and their locations in the galaxy are strongly correlated. The outer, or galactic halo, part has only very old stars, with less than about 10% of the solar metal abundance. These are called Population II stars, and they are not likely to have Earth-like planets. The part of the galactic disk near the Sun (see Figure 5.8) has stars with a mix of ages, masses, and compositions. The youngest ones, with close to solar metallicity, are called Population I stars and have probably not been around long enough for life to evolve on their planets (though there is hope for the future). When you look outward, to the part of the disk farther from the

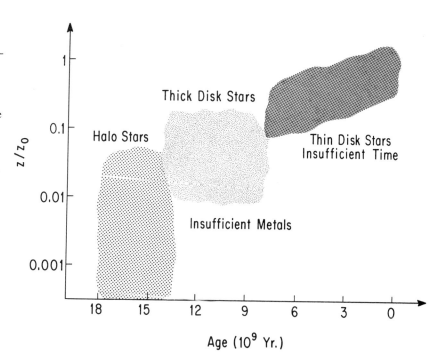

FIGURE 5.8

The age-metallicity relation for stars in the solar neighborhood. The vertical axis gives metal abundance, Z, in units of the metal abundance Z_0 in the Sun. Most halo and thick-disk stars are probably too deficient in heavy elements to have formed terrestrial planets, while most of the metal-rich, thin-disk stars are younger than the Sun, allowing insufficient time for life to develop. Nearer the center of our Milky Way galaxy is another population, called the bulge, which is both relatively old and relatively metal-rich. This is the most promising site for life bearing planets.

galactic center than we are, you see younger stars and fewer metals. Looking inward toward the center, you find that the stars are, on average, both older and more metal-rich.

It seems to me likely that lots of **habitable planets** are available in the Milky Way, but that most of them will be a good deal closer to the galactic center than we are. The round-trip time for a radio (or light) signal to travel from us to the galactic center is about 50,000 years. Thus, while inhabited planets may be common, distance will greatly increase the difficulties of communicating with their hypothetical citizens, whether Republican or Democrat.

REFERENCES

Bethe, H. A. 1939. Energy production in stars. *Physical Review* 55: 434–456.

Bethe, H. A., and Critchfield, C. L. 1938. The formation of deuterons by proton combination. *Physical Review* 54: 248–254.

Burbidge, E. M., Burbidge, G. R., Fowler, W. A., and Hoyle, F. 1957. Synthesis of the elements in stars. *Reviews of Modern Physics* 29: 547–650.

Cameron, A.G.W. 1957. Chalk River Report CRL-41 and Nuclear reactions in stars and nucleogenesis. *Publications of the Astronomical Society of the Pacific* 69: 201–222.

Chandrasekhar, S. 1931. The maximum mass of ideal white dwarfs. *Astrophysical Journal* 74: 81–82.

Chandrasekhar, S. 1935. The highly collapsed configurations of a stellar mass. *Monthly Notices of the Royal Astronomical Society* 95: 207–225.

Gamow, G. 1949. On relativistic cosmogony. *Reviews of Modern Physics* 21: 367–373.

Merrill, P. 1952. Technetium in the stars. *Science* 115: 484.

Payne, C. H. 1925. *Stellar Atmospheres* (Cambridge, UK: Heffer & Sons).

Tinsley, B. M. 1968. Evolution of the stars and gas in galaxies. *Astrophysical Journal 151*: 547–565.

Tinsley, B. M. 1980. Evolution of the stars and gas in galaxies. *Fundamentals of Cosmic Physics 5*: 287–388.

Trimble, V. 1975. Origin and abundances of the chemical elements. *Reviews of Modern Physics 47*: 877–976.

Trimble, V. 1991. Origin and abundances of the chemical elements revisited. *Astronomy and Astrophysics Reviews 3*: 1–46.

CHAPTER 6

THE ORIGIN AND EVOLUTION OF LIFE IN THE UNIVERSE

Christopher P. McKay

INTRODUCTION

Are we alone? We wonder whether life on Earth is a rare, even unique, phenomenon or whether life is to be found in myriad forms on many worlds throughout the Universe. As human beings, we ponder the probability that life elsewhere has developed intelligence to levels comparable with ours; and if so, how can we communicate with these life-forms? The question of life in the Universe necessarily straddles scientific fields from astronomy to zoology — fields that appear at first to have no common thread.

The nature of the Universe we can observe, from the perspective of life on the Earth, suggests that indeed there are numerous planets in our own galaxy and that life could be found on virtually all that are Earth-like in their size and distance from a central star. The first test of our theories for the distribution of life on other planets will be a search for past life on Mars. The question of advanced life and, in particular, intelligent life elsewhere is more difficult to address and the only strategy for presently seeking intelligent life is to listen for radio broadcasts from alien transmitters.

Life begins at home, and so this chapter surveys the origin and evolution of life by reviewing what we have learned from life on Earth. From this single point of data one can extrapolate outward to the stars.

LIFE ON EARTH

Despite its various forms, there is only one life on Earth. One of the most profound results of modern science is the unity of biochemistry. All life-forms on Earth have the same basic biochemical makeup. This fact is perhaps best illustrated by the **amino**

acids from which all **proteins** are made. Of all possible amino acids, life on Earth is based primarily on a common set of 20. These are shown in Figure 6.1 along with the other basic biochemicals of life (Lehninger, 1975). The amino acids in Figure 6.1 come in left- and right-hand versions. The handedness refers to the relative location of the molecular groups. In nonbiological processes, amino acids are found equally in the left- and right-hand versions; in biology, only the left-hand amino acids are used. Another universal characteristic of all life is that it shares the same genetic code — **RNA** and **DNA**. Life is a particular set of biochemicals, and all life on Earth, from the *Escherichia coli* bacteria to the blue whale, is constructed from that same set.

The similarity between the genetic material of all life implies another important property of life on Earth; it shares a common ancestor. Thus, it is possible to construct a family tree — known as a **phylogenetic tree** — that shows how all life-forms are related. One such tree, based on comparing RNA, is shown in Figure 6.2 (Woese, 1987). Other trees can be constructed with the location of the eukarya being the main variant (Rivera and Lake, 1992). The tree of life shown in Figure 6.2 divides into three main branches: the eukarya, the bacteria, and the archaea. The eukarya are virtually all breathers of oxygen and include all higher plants and animals. The eukarya comprise what we refer to as life in the common use of the term. Multicellular life is found only in the eukarya.

The other two branches of the tree of life contain the microscopic life-forms commonly described as bacteria. Technically these forms are called *prokaryotes*. They differ from the eukarya in that they do not have nuclei in their cells; their genetic material is distributed throughout the intracellular medium. Studies of the tree of life shown in Figure 6.2 indicate that the common ancestor of all life on Earth today, the root of the phylogenetic tree, was a prokaryote that lived in hot water and metabolized sulphur (Woese, 1987). Similar organisms can be seen today in sulfurous hot springs, such as those in Yellowstone Park. The common ancestor was not necessarily the first cell. To the contrary, the common ancestor almost certainly represented a considerable period of evolution after the origin of life.

There are three possible explanations why the common ancestor of life inhabited a sulfurous hot spring. One possibility is that such an environment may have been crucial to the origin of life or to its early evolution. Alternatively, early life on Earth may have been stressed by some ecological disaster, such as a large comet impact, that exterminated all life except some microorganisms living deep within a subsurface hot spring. After the event, these organisms recolonized the planet. The nature of the common ancestor may also be purely a result of chance and be unconnected in any fundamental way to the events on the early Earth.

The origin of life on Earth remains a puzzle. A timeline (see Figure 6.3) shows that life appears early in Earth's history. The formation of the Earth was completed about 3.8 billion years ago when the last of the **planetesimals** that formed the planets were accreted. This event, known as the **heavy bombardment**, is recorded in the densely cratered surfaces of the Moon, Mercury, and Mars. On Earth the craters caused by the heavy bombardment have long been eroded.

In the **fossil record** there is definite evidence for life at 3.5 billion years ago (Schopf, 1993). Furthermore, the nature of this life indicates that oxygen-generating **photosynthesis** was present and formed the basis of complete microbial communities similar to **microbial mats** found in many locations on Earth today (Schopf, 1993). Thus, there was certainly comparatively sophisticated microbial life by 3.5 billion years ago. Moreover, there is indirect evidence that life on Earth may have existed even

FIGURE 6.1

The Primordial Biomolecules: The basic building blocks of life on Earth. (Reprinted by permission from Lehninger, 1975.)

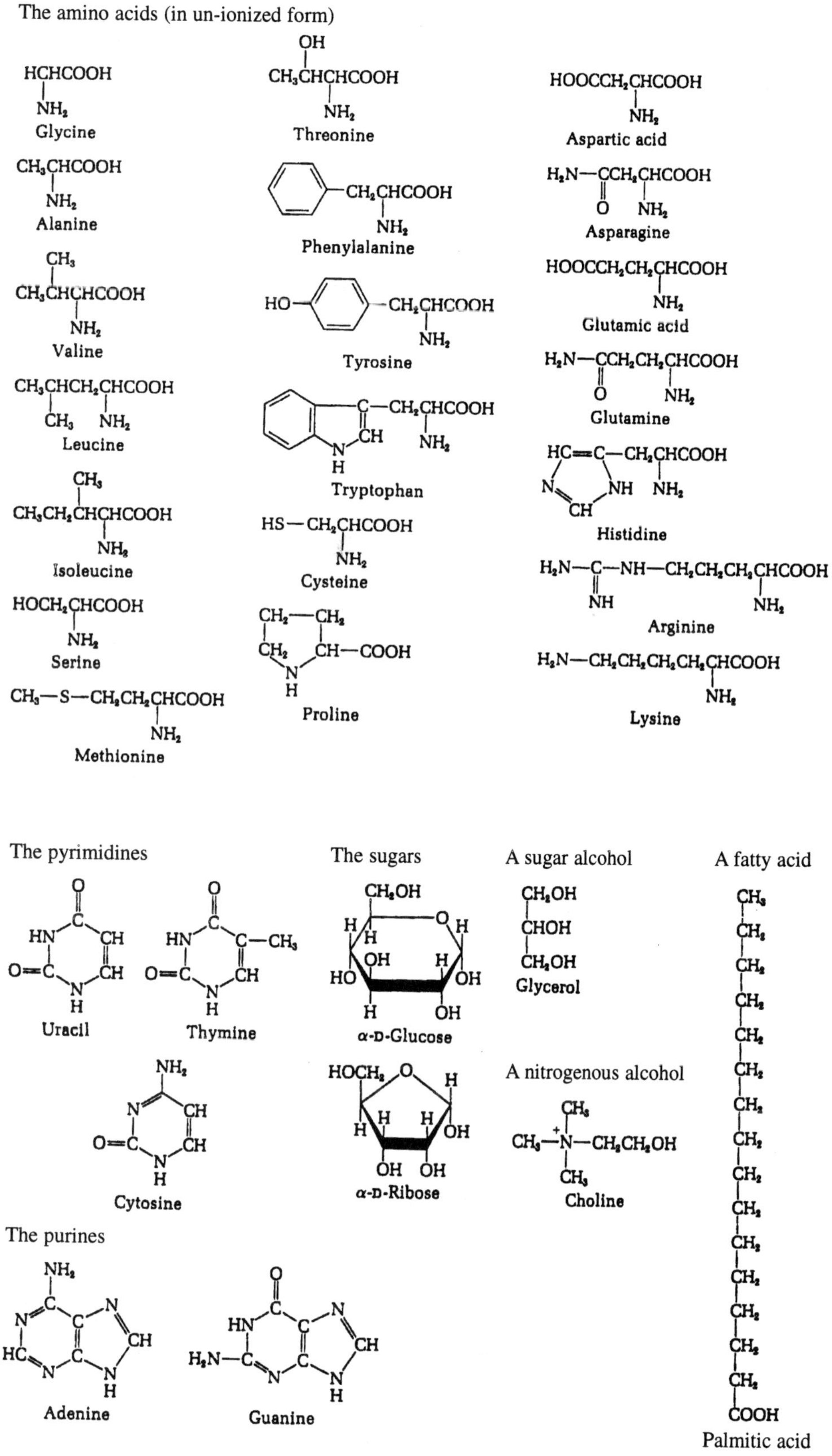

The amino acids (in un-ionized form): Glycine, Alanine, Valine, Leucine, Isoleucine, Serine, Methionine, Threonine, Phenylalanine, Tyrosine, Tryptophan, Cysteine, Proline, Aspartic acid, Asparagine, Glutamic acid, Glutamine, Histidine, Arginine, Lysine.

The pyrimidines: Uracil, Thymine, Cytosine.

The purines: Adenine, Guanine.

The sugars: α-D-Glucose, α-D-Ribose.

A sugar alcohol: Glycerol.

A nitrogenous alcohol: Choline.

A fatty acid: Palmitic acid.

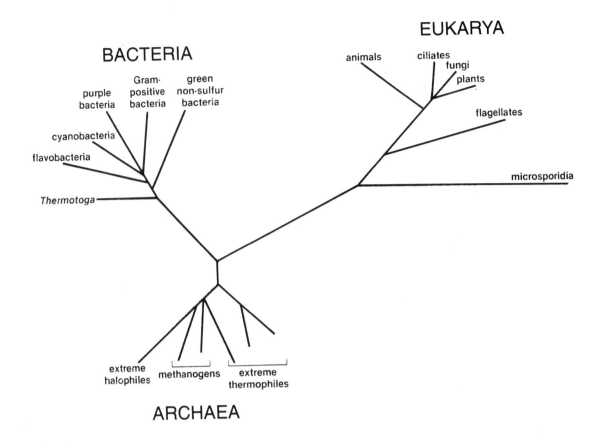

FIGURE 6.2

Tree of life showing the genetic relationship between all life forms on Earth. The eucarya include most of what is commonly referred to as life: plants, animals, fungi, and so forth. The archaea and the bacteria are commonly referred to as bacterial. All life is related and has a common ancestor that lived in a hot springs environment and metabolized sulfur. (Reprinted by permission from Woese, 1987.)

earlier. Life, photosynthesis in particular, produces a shift in the relative amounts of the **isotopes** of carbon. Carbon within living systems is about 2% enriched with the lighter isotope of carbon, ^{12}C, compared with the heavier isotope, ^{13}C. This shift is seen in sedimentary material of biological origin for the past 3.5 billion years. Interestingly, this shift is also seen in **sedimentary deposits** that are 3.8 billion years old (Schidlowski, 1988) — during the end of the heavy bombardment. This information is consistent with photosynthetic life at that time, but it is not proof of life because the sediments are too altered to preserve any direct fossils. Nonetheless, the evidence from Earth indicates that life arose very rapidly and possibly instantaneously (which, geologically speaking, is a period of approximately several tens of millions of years). A clear prediction from this observation is that on any planet with conditions similar to those on early Earth, life would quickly emerge, even as the planet was forming.

Although we know that life appeared early in the history of Earth, we do not know the details of how life originated. The most widely held theory for the origin of life is based on the original suggestions of A. I. Oparin and J. B. S. Haldane. Independently,

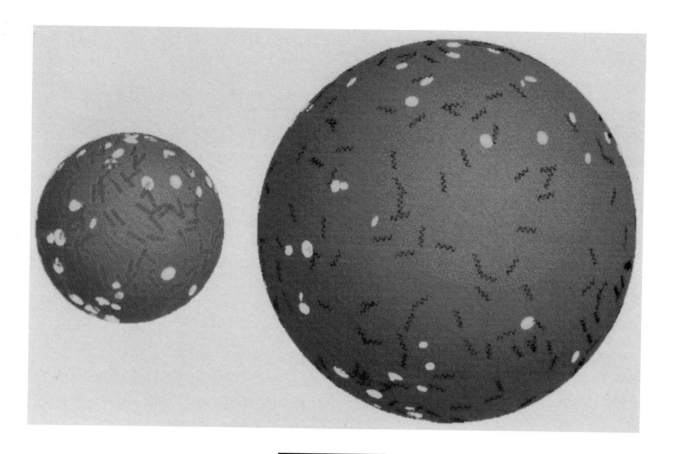

Color Plate 1.1

Expanding balloon analogy for the Universe. Note that the galaxies (yellow blobs) stay the same size, but the distances between the galaxies all grow by the same factor. Photons move through the Universe and get redshifted as the Universe expands.

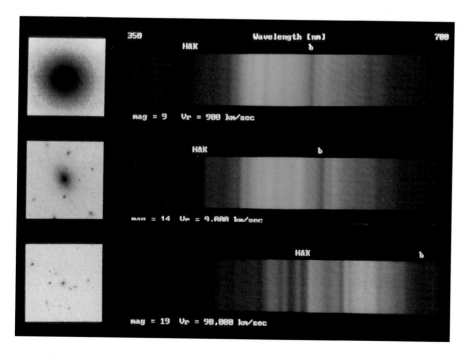

Color Plate 1.2

Spectra and images of nearby and distant galaxies. Note how the dark H&K lines produced by calcium ions and the b lines produced by magnesium atoms shift toward the red end of the sprectrum, the more so the larger is the recessional velocity, v, of a galaxy.

Equal Power on All Scales Model

COBE Data

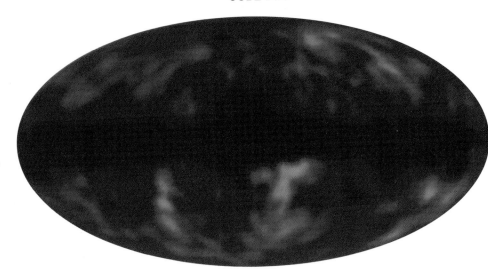

Color Plate 1.4

Top: Model sky map based on equal power on all scales.

Bottom: Actual sky map seen by COBE. The horizontal red bar in the COBE data is microwave emission from the Milky Way galaxy.

Color Plate 1.5

Infrared picture of the sky at 100-μm wavelength from the COBE satellite. The "lazy S" is emission from our Solar System, while the bright horizontal bar is emission from our Milky Way galaxy.

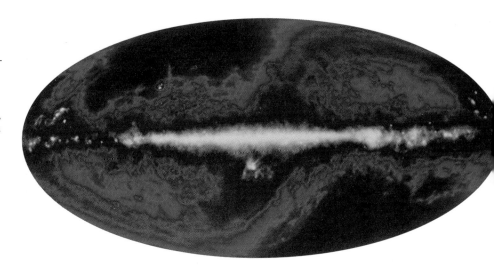

Color Plate 3.1

Optical and infrared photographs of the star-forming region in the constellation of Orion. The optical picture (*left*) shows relatively few stars. The infrared picture (*right*) shows a large number of objects. Because infrared light can more easily escape the molecular cloud, the comparison of these two pictures illustrates that the young stars are deeply embedded within the cloud. (The optical image is from the 4-m telescope at Kitt Peak National Observatory and the infrared picture was obtained by Dr. Ian Gatley and colleagues with the Kitt Peak 1.3-m telescope.)

Color Plate 4.1

The Helix nebula (NGC 7293) in Aquarius, a well-known planetary nebula about 400 light-years away. (© Anglo-Australian Observatory. Photograph by David Malin.)

Color Plate 4.2

SN 1987A near the Tarantula nebula in the Large Magellanic Cloud. The supernova is the bright star in the lower left corner of the picture. (© Anglo-Australian Observatory. Photograph by David Malin.)

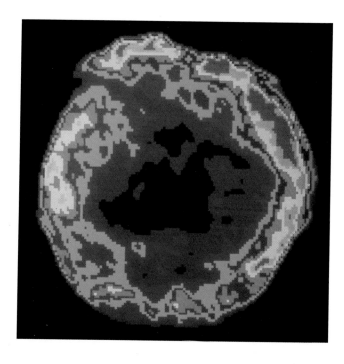

Color Plate 4.3

Remnant of Tycho's supernova of 1572, imaged at radio wavelengths. False-color mapping is used to depict different values of the surface brightness. (Courtesy National Radio Astronomy Observatory/Associated Universities, Inc.)

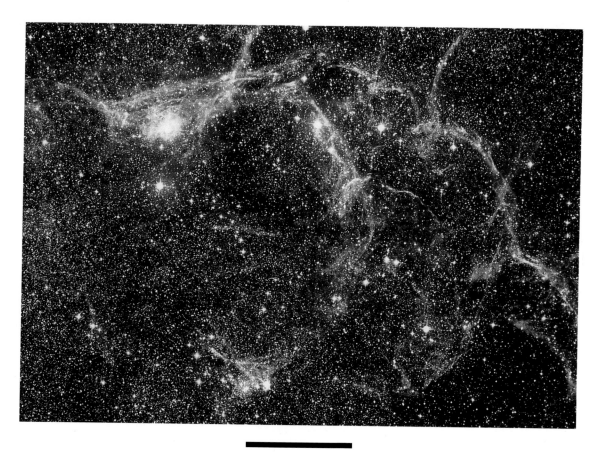

Color Plate 4.4

The Vela supernova remnant, 1500 light-years away. Its age is estimated to be about 20,000 years. (© Royal Observatory Edinburgh/Anglo-Australian Observatory. Photograph made from UK Schmidt plates by David Malin.)

Color Plate 4.5

Left: The central region of the Orion nebula (M42), in which stars have recently formed. The nebula is about 1600 light-years away. (Courtesy Lick Observatory.)

Color Plate 4.6

Below: M83, a spiral galaxy in the constellation Hydra. (© Anglo-Australian Observatory. Photograph by David Malin.)

Color Plate 4.7

Below: Type Ia supernova 1994D in NGC 4526, discovered by the author and collaborators on March 7, 1994, with a 0.8-m robotic telescope at Leuschner Observatory. The supernova, roughly 50 million light-years away, is the brighter of the two visible stars; the fainter star to the left of the supernova is in our own Milky Way galaxy. The dark streak is produced by dust in NGC 4526.

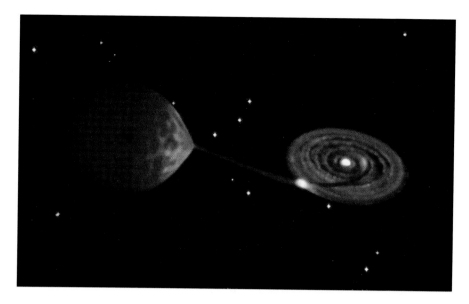

Color Plate 4.10

Accretion of matter by a compact star (white dwarf, neutron star, or black hole) from its companion in a binary system.

Color Plate 4.11

Interior of a red supergiant immediately before the collapse of its iron core and subsequent explosion as a Type II supernova. The relative sizes of the inner layers are not shown to scale. (Diagram by Jose R. Diaz. © 1993 Sky Publishing Corporation. Reproduced with permission.)

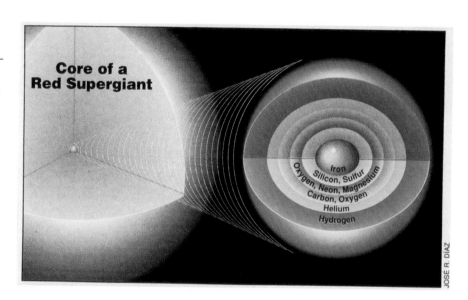

Color Plate 4.12

The Milky Way (*near bottom*) and the Magellanic Clouds, two small satellite galaxies. (Courtesy European Southern Observatory.)

Color Plate 4.13

"Before" and "after" photographs of SN 1987A. (© Anglo-Australian Observatory. Photograph by David Malin.)

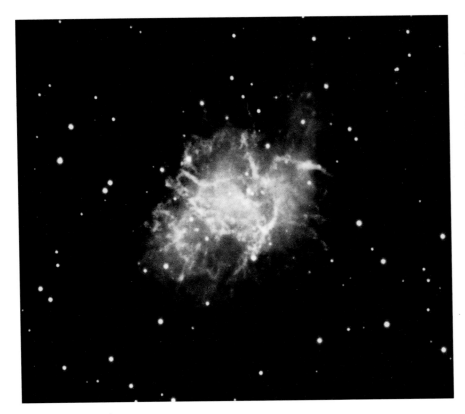

Color Plate 4.16

The Crab nebula (M1) in the constellation Taurus, the remnant of the supernova of 1054 AD. (Courtesy Lick Observatory.)

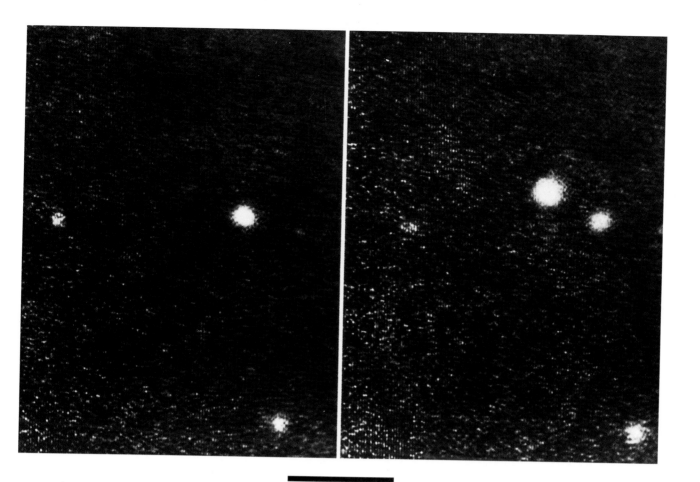

Color Plate 4.17

Photographs of the Crab pulsar when it is pointing toward us (*right*) and away from us (*left*). (Courtesy J. S. Miller and E. J. Wampler, Lick Observatory.)

Color Plate 4.18

Hubble Space Telescope image of SN 1987A obtained in February 1994, after the telescope refurbishment mission. The supernova ejecta are barely resolved in the *center*. Three rings of gas are visible, probably associated with material expelled by the progenitor star before the supernova explosion. (Courtesy NASA/Space Telescope Science Institute.)

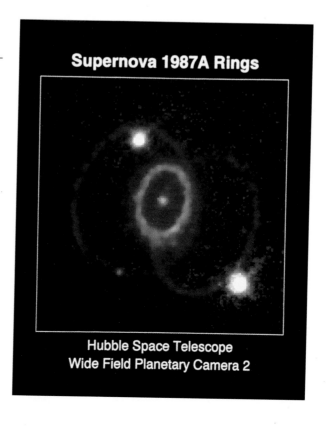

Color Plate 6.6

Lake Vanda in Wright Valley, one of the dry valleys in Southern Victoria Land, Antarctica. Lake Vanda is about 5-km long. Although the mean temperature of the Antarctic dry valleys is 20 degrees below freezing, deep lakes (>30 m) are formed by transitory melting of glacial ice; the same could have existed on Mars and provided a refuge for life as that planet cooled. (Photo by Robie Vestal.)

Color Plate 6.7

A: Microbial mat on the bottom of an ice-covered Antarctic lake (image is about 1-m across). (Photo by Dale Andersen.)

B: NASA scientist preparing to enter an ice-covered lake. (Photo by Dean Schwindler.)

Color Plate 6.8

Life within the porous sandstone in the highlands of the Antarctic dry valleys, where lichens grow in subsurface pore spaces (scale about 1 cm from top of rock to bottom of figure). (Photo by I. Friedmann.)

Color Plate 6.9

Viking orbiter image of Hebes Chasma (0°S, 75°W), a box canyon about 280-km long. The mesa in the center of the canyon shows layered sediments that are believed to have been deposited in standing bodies of water.

Color Plate 6.12

SPOCK: It's life, Jim, but *not as we know it*!

KIRK: Set phasers to "Stun"!

(From "The Devil in the Dark" episode of Star Trek, Paramount/NBC. © 1967, Motion Picture & Television Photo Archive. All rights reserved. Used with permission.)

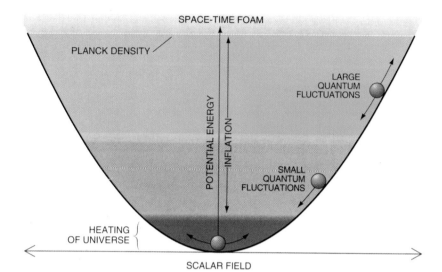

Color Plate 7.1

Scalar field in an inflationary Universe can be modeled as a ball rolling down the side of a bowl. The rim corresponds to the Planck density of the Universe, above which lies a space-time foam, a region of strong quantum fluctuations of space-time. Below the rim (green), the fluctuations are weaker, but may still ensure the self-reproduction of the Universe. If the ball stays in the bowl, it moves into a less energetic region (orange), where it slides down very slowly. Inflation ends once the ball nears the energy minimum (purple), where it wobbles around and heats the Universe.

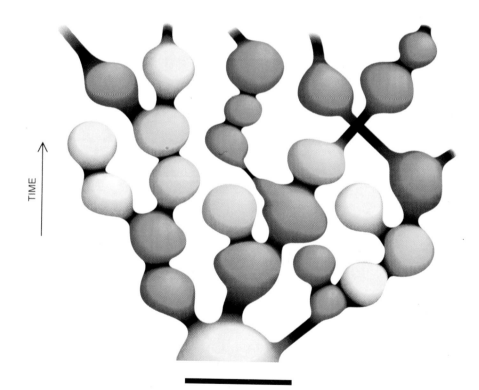

Color Plate 7.3

Self-reproducing cosmos appears as an extended branching of inflationary bubbles. Changes in color represent mutations in the laws of physics from parent universes. The properties of space in each bubble do not depend on the time when the bubble formed. In this sense, the Universe as a whole may be stationary, even though the interior of each bubble is well described by the Big Bang theory.

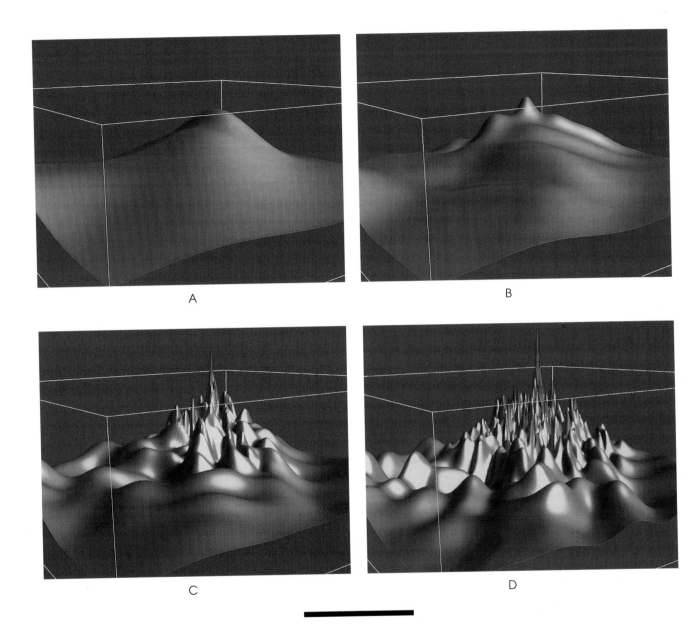

Color Plate 7.4

Evolution of the scalar field ϕ during inflation. Quantum fluctuations create peaks of the distribution of the scalar field. In the vicinity of each such peak, the Universe expands with enormous speed, which leads to creation of new inflating regions of the Universe. This is the physical mechanism of the self-reproduction of an inflationary Universe.

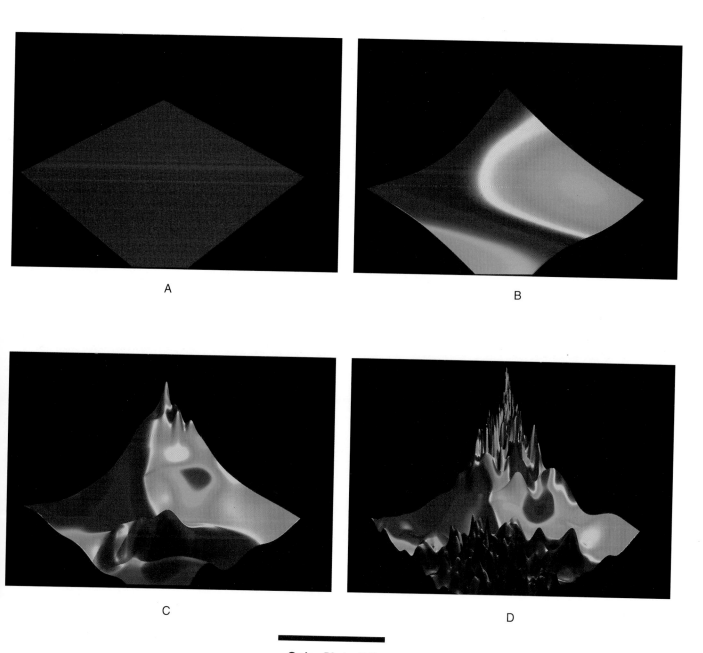

Color Plate 7.5

Evolution of the scalar fields φ (height of the distribution) and Φ (color) during inflation. Fluctuations of color imply changes of low-energy laws of physics in different parts of the Universe.

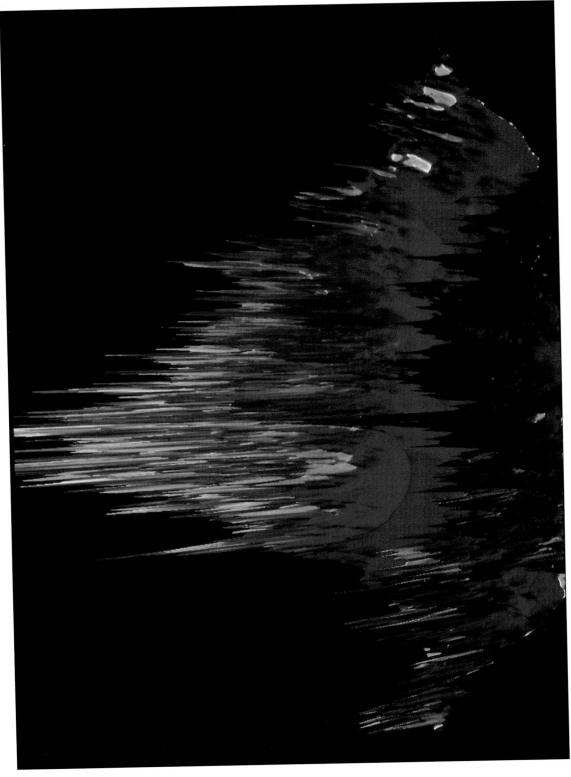

Color Plate 7.6

After a sufficiently long time, the inflationary Universe becomes divided into many exponentially large regions with different laws of low-energy physics in each region. The laws are fixed in the valleys, where the energy density (the value of the scalar field φ) is small, but they are rapidly fluctuating near the peaks of the mountains. Each such peak, for practical purposes, can be considered as a beginning of a new Big Bang.

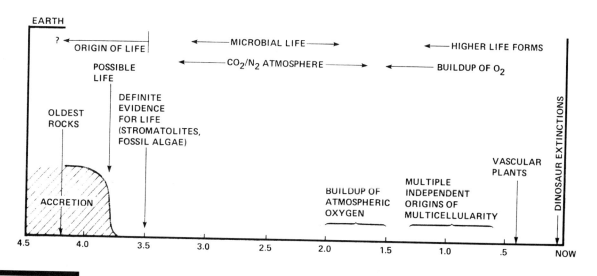

FIGURE 6.3

Timeline in billions of years for planet Earth. Life evolves quite rapidly after the end of the formation of Earth, 3.8 billion years ago. The next major event in the history of life is the buildup of oxygen, starting about 2 billion years ago, followed by, and possibly causing, the rise of multicellular life, about 0.6 billion years ago. (The last major event in the history of life, the development of intelligence, occurred so recently that it cannot be resolved on this plot.) (Adapted from McKay and Stoker, 1989.)

they suggested that life originated from **organic material** that had been produced abiologically. This theory for the origin of life received convincing support in the experiments of Stanley Miller. In these experiments, and in subsequent work along these lines, gas mixtures thought to represent the plausible composition of the atmosphere of early Earth were subjected to electrical discharges (Miller, 1992). The design of Miller's apparatus is shown in Figure 6.4.

These experiments are remarkable in two ways. First, the fact that organic molecules are so readily made from inorganic, albeit **reducing**, gases suggests that organic chemistry, far from being the sole purview of living systems, should be commonplace in the cosmos. Second, the organics produced in the **abiotic synthesis** were not a mere random collection of molecules but were composed of many of the same compounds that are found in life. The chemistry of life appears, therefore, to be part of cosmic organic chemistry.

This overall view of cosmic organic chemistry received considerable support by the subsequent discovery of organics in space. Organics were first discovered in meteorites and the organics present were similar to those produced in Miller's abiotic syntheses. Organics were then discovered in the interstellar medium (Irvine and Knacke, 1989), in comets (Kissel and Krueger, 1987), in the outer solar system (Encrenaz, 1984), and, especially, on Titan (Sagan et al., 1984). In fact, only the absence of organics on Mars (Biemann et al., 1977) is incongruous with a prevailing organic chemistry in the solar system and the cosmos.

Although the evidence for the widespread abiotic origin of organics relevant to biochemistry is compelling, the standard theory for the origin of life suffers from a major flaw — it has not been possible to create life from abiotic organics. This is a more serious issue than one might imagine at first. Lazcano and Miller (1994) have recently

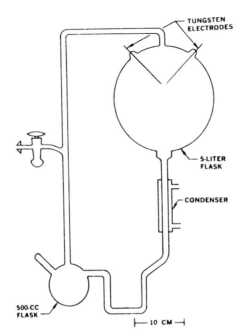

FIGURE 6.4

Diagram of the apparatus used by Miller demonstrating the nonbiological production of organic molecules similar to those in life. The flask is filled with a mixture of NH_3, CH_4, and H_2O; an electric discharge is passed through the electrodes. Organic molecules, including amino acids, are produced. This experiment supports the standard theory for the origin of life in which the first organism arises from and is nourished by a primordial soup of organics produced by nonbiological processes. (Adapted from Miller, 1992.)

suggested that the time required for the emergence of life from the prebiotic soup may have been short (on geological timescales) — less than 10 million years. If this is true, a factor of 100 million enhancement in the experimental process compared with the process in nature should produce life in the laboratory in 1 year. When one compares random processes in nature with directly controlled experiments, it is not unreasonable to expect that rate enhancements of 100 million are possible. Thus, it is problematical that, 40 years after the first experiment, life remains to be created in abiotic simulations.

There are alternate theories for the origin of life (Davis and McKay, 1996). Some postulate that life came to Earth from elsewhere. This notion, called **panspermia**, has been gaining attention as researchers appreciate the short interval for the origin of life on Earth and as mounting evidence shows how impacts can launch materials from one planet to another in a way that could preserve dormant spores (Melosh, 1988). Other theories for the origin of life allow a terrestrial origin but differ from the standard theory in that they do not postulate an organic origin for life. Instead, they postulate that the first life-forms were composed of clay minerals that evolved into organic-based life later (Cairns-Smith, 1982). Although the standard theory for the origin of life assumes that the first life-forms lived by consuming organic material already present in the environment, a process known as *heterotrophy*, other theories suggest that the first life-form could have derived its energy from sunlight or even from chemical reactions (such as $4H_2 + CO_2 \rightarrow CH_4 + 2H_2O$). Even if life did arise from a **primordial soup** and stayed alive by consuming that soup, the question of where the ingredients for the soup came from still remains.

The original work of Miller and others suggested that the organics in the soup were produced on Earth in an atmosphere rich in methane and ammonia — considerably different from our present atmosphere. However, methane and ammonia may not have been stable in Earth's early atmosphere because of destruction by **ultraviolet** (UV) light from the Sun, which has led to the suggestion that the ingredients for the prebiotic soup were carried to a receptive Earth by comets and meteorites (Oró, 1961). One common feature of all theories for the origin of life on Earth is that life required liquid water environments (Davis and McKay, 1996). The diversity of theories for the origin of life illustrates how uncertain we are about the events that led to life on Earth. What we can say for certain is that the origin of life on Earth occurred rapidly and required liquid water.

An interesting biochemical development that bears on the origin and early development of life is the discovery that RNA can act both to store genetic information (such as DNA) and to affect biochemical reactions (such as enzymes, which are proteins). In modern life-forms, genetic information is recorded in DNA but, before the information can be used, it must be transcribed into proteins, the working molecules of biochemistry. The separation of biochemistry into information molecules and action molecules was long a puzzle. Because the molecules were mutually interdependent, which originated first was unclear. The discovery that RNA is simultaneously capable of both functions resolved the dilemma.

An early stage in the evolution of life may have involved a so-called RNA world composed of organisms in which RNA was the basis for life and this remarkable molecule was both gene and enzyme (Joyce, 1989). Later evolution led to development of more efficient genetic material (DNA) and more efficient enzymes (proteins) and formed the basis for the DNA world we have now. The timing of this evolution is uncertain but it would appear to have occurred before the oldest fossils, at 3.5 billion years ago. The DNA world must have occurred before the divergence of life into the three main groups shown in Figure 6.2 because comparison of the genetic difference between these groups of life implies that the genetic code is 3.8 ± 0.6 billion years old (Eigen et al., 1989).

Although its origin is shrouded in mystery, life has left a more complete account of its subsequent history in the geological record. We know that life on Earth remained microbial for about 3 billion years. For about the first half of this time period, the atmosphere of Earth did not contain any appreciable oxygen. Although, almost certainly, photosynthetic microorganisms were producing oxygen, it was consumed by natural sources of reducing material, including H_2 from volcanos, dissolved iron in the oceans, and organic material. A net production of oxygen is achieved only when the organic matter produced in photosynthesis is secured in sediments. Exposed organic material oxidizes and thus consumes the O_2 that was created when the organic matter was produced. Not until about 2 billion years ago did oxygen production overcome these losses so that oxygen could begin to accumulate in the atmosphere. The rate of buildup of atmospheric oxygen is uncertain, but probably continued until 600 million years ago.

After the oxygen buildup, multicellular life that may have been linked to the level of atmospheric oxygen rapidly developed. Oxygen may have been critical to the development of multicellular life in two ways. First an ozone layer, formed after oxygen accumulated, may have been needed to protect multicellular life from harmful UV light. Second, and more probable, oxygen may have been needed to provide a metabolic reaction powerful enough to supply the needs of large multicellular organisms. All multicellular life-forms breathe oxygen.

The buildup of oxygen on Earth is one of the key events in the history of life on this planet. Oxygen production was dominated by biological photosynthesis, but geological processes controlled the rate at which organic material was sequestered and the rate at which reduced gases were vented to the atmosphere (DesMarais et al., 1992). Thus, the timing of an oxygen-rich atmosphere and the concomitant development of multicellular life was set by the geological nature of Earth (Knoll, 1985).

Ultimately, the development of multicellular life led to the emergence of intelligent life on Earth. The evolutionary path was not a straight one; there were considerable periods of time during which intelligence at the level of humans could have evolved but did not. Perhaps the most notable of such periods was the dinosaur era. If these creatures were as biochemically and behaviorally sophisticated as is now thought, the fact that they did not develop an intelligent species during the 150-million-year period in which they dominated Earth is puzzling (McKay, 1995). It has been suggested that dinosaur species that could have been on the road to intelligence existed as long as 12 million years before the demise of dinosaurs 65 million years ago (Russell and Séguin, 1982). In this context, it is interesting to note that humans developed from simple primates in less than 3 million years.

The development of human-like species can likely be reduced to three essential steps: 1) the origin of life, a rapid, biochemical event; 2) the rise of oxygen and multicellularity, a slow, geologically determined event; and 3) the development of intelligence, a random, possibly rare, event. These considerations suggest that human levels of intelligence may be a rare phenomenon in the cosmos but simpler life-forms may be abundant (McKay, 1995).

Microbial life on Earth has adapted to many unusual conditions, and the distribution of microorganisms tells us the factors that are truly limiting for life. Microorganisms can thrive in boiling water, in acid, and in concentrated salt solutions. Other microorganisms live on the bottom of the sea floor in **hydrothermal vents** where they consume H_2S by reacting with O_2 dissolved in seawater (the ultimate source of this O_2 is photosynthesis on the surface of Earth). Many microbes live completely without O_2, and, in fact, oxygen is a deadly poison to them. These anaerobic bacteria are responsible for the degradation of waste material in landfills and sewage. However, the one environmental requirement that all life on Earth shares is the requirement for liquid water. Some organisms can survive dormant in a dried state, but to grow all life on Earth needs liquid water (Kushner, 1981). Ice will not do. Life grows in ice and snow only when there is a liquid fraction. Vapor will not do, although some organisms, notably **lichens**, can concentrate vapor to liquid as long as the **relative humidity** is above 70%. In addition to liquid water, the requirements for life are energy, carbon, and some other elements listed in Table 6.1 along with an indication of how commonly these requirements are satisfied on the other planets of our solar system.

TABLE 6.1 **Requirements for Life**

Requirement	Occurrence in Our Solar System
Energy (usually sunlight)	Common
Carbon	Common, as CO_2 and CH_4
Liquid water	Rare
Nitrogen, phosphorus, sulfur, and other elements	Common

Source: Adapted from McKay, 1991.

LIFE ON MARS

Liquid water is the defining and quintessential requirement for life on Earth. Therefore, the search for liquid water is an operational approach to the search for life elsewhere. An Earth-like planet is, by definition, a planet that contains liquid water; by analogy with early Earth, any such planet should quickly see the emergence of life.

In our own solar system, only one instance of evidence for liquid water exists beyond Earth. Liquid water flowed on Mars repeatedly during the past several billion years. The orbital images showing fluvial features are a direct indication that a liquid — water is the only plausible candidate — flowed on the Martian surface (see Figure 6.5).

In 1976, the Viking spacecraft landed on Mars and began a search for microscopic life-forms in the sands. The Viking biology package was composed of three instruments, each of which conducted experiments on a sample of the Martian surface. One experiment searched for photosynthesis, and the other two looked for organisms capable of consuming a nutrient solution — a soup-like broth. The results were intriguing. In both experiments with the nutrient solution, gases were released, which indicated some sort of activity. The results were exciting and the prospect that life had been discovered loomed large until the results of another instrument were considered. Each of the two Viking landers also carried a sophisticated instrument for the characterization of organic material, a **gas chromatograph mass spectrometer** (GCMS). The GCMS did not detect organics in the Martian samples at the level of parts per billion. It detected only some cleaning agents used on the spacecraft before launch. The prospect of life without organic material seemed so unlikely that most researchers attribute the reactivity seen in the Viking biology experiments to some unusual sort of surface chemical reactivity. The mostly likely source of the chemical reactivity is one or more **oxidants**, such as hydrogen peroxide, produced by the strong solar UV light reaching the Martian surface (Zent and McKay, 1994). Interesting, but hardly life.

Why is Mars so dead? The fundamental answer must surely lie in the fact that under present conditions there is never liquid water on the surface of Mars at any place or at any time of the year. The absence of liquid water, exacerbated by the oxidizing surface, the UV light, and the thin atmosphere (less than 1% as thick as Earth's), makes today's Mars as dead as a doornail. However, because there is direct evidence that Mars had liquid water flowing on its surface, there should be other evidence that Mars once had a more active and habitable climate. Indeed, the large volcanos, such as Olympus Mons, that now appear to be extinct are mute testimony to a once geologically active planet.

Geomorphological evidence suggests that liquid water existed on the surface of Mars at approximately the time that the first life appeared on Earth, between 3.8 and 3.5 billion years ago (McKay and Stoker, 1989). The possibility of the origin of life on Mars is based on analogy with Earth. All the major habitats and microenvironments that would have existed on Earth during the formation of life would have been expected on early Mars as well: hot springs, salt pools, rivers, lakes, volcanos, and so forth. Even tidal pools would have existed on Mars, albeit at a much reduced level because there would have been only solar tides. The possible nonbiological sources of organic material would have supplied both planets. Perhaps the major unknown is the duration of time that Mars had Earth-like environments compared with the time required for the origin of life. The length of neither of these times is known precisely, but current theories suggest that the lengths may be comparable (McKay and Davis, 1991).

FIGURE 6.5

Orbiter image of fluvial channels on Mars, which shows that at one time liquid water flowed across the surface of Mars. The presence of liquid water on Mars is the primary reason that a search for evidence of past life is being considered. This site is Warrango Vallis, located on the ancient cratered terrain on Mars (48°S, 98°W). The extent of the scene is about 200 km. The source of the water may have been precipitation or subsurface melting. (Viking frame 63A09.)

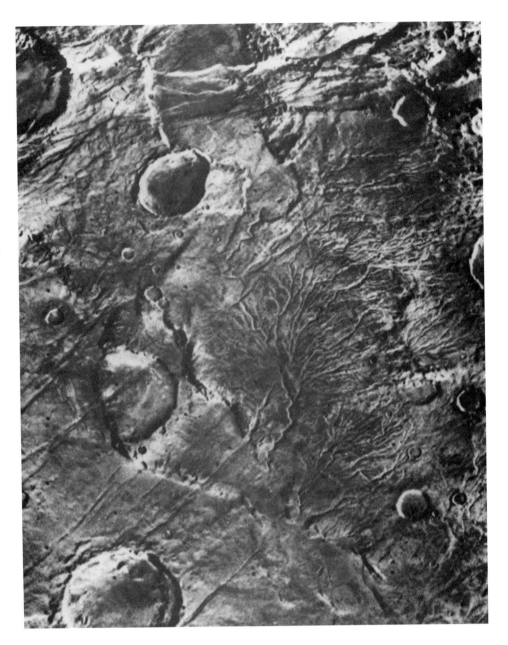

The events that led to the origin of life on early Earth and that may have also occurred on early Mars may be better preserved in the sediments on Mars than those on Earth. On Earth, sediments that date back 3.5 to 4.0 billion years ago are rare, and those that exist are usually severely altered. On Mars, over half the planet dates back to the end of heavy bombardment, about 3.8 billion years ago, and has been well-preserved at low temperatures and pressures. Thus, although there may be no life on Mars today, Mars may hold the best record of the events that led to the origin of life on Earth-like planets.

ANTARCTICA AND MARS

To understand how life might have survived on Mars as the planet slowly froze we turn to the coldest, driest place on Earth — the dry valleys of Antarctica (McKay, 1993). These valleys are extreme deserts, with mean annual temperatures of −20°C and an annual precipitation equal to a layer of water about 2 cm thick — drier than most hot deserts. In the summer, under continuous sunlight, the temperatures just barely climb above freezing on the valley floors for a few days (Clow et al., 1988). Conditions in these Mars-like valleys are too dry for life and they appear lifeless. However, two ecosystems are hidden in the dry valleys, and life has learned to thrive on the small, transient amounts of liquid water that occur in the summer months.

On the floor of the dry valleys are located perennially ice-covered lakes (see Color Plate 6.6). Inside these lakes, growing on the feeble light that penetrates the constant ice cover, is a lawn of algae (see Color Plate 6.7). The persistence of lakes in an environment in which the mean temperature is 20 degrees below freezing is paradoxical. What allows for such lakes is the fact that during most summers air temperatures climb above freezing for a few days. During this time, the glaciers that surround the valleys begin to melt and the meltwater flows into the lakes. The energy carried into the lake by the meltwater is trapped by the ice cover. The rate at which this energy can be conducted through the ice cover determines the thickness of the ice (McKay et al., 1985). For the dry valleys, this results in about 4 m of ice overlying tens of meters of liquid water. In this sense, the dry valleys are not so dry. About 1% of the sunlight used for photosynthesis that is incident on the lake surface is able to penetrate the thick ice cover (McKay et al., 1994). This is enough to support a thriving assemblage of algae and bacteria that live on the bottom of the lake. There are no fish or higher plants in the lakes.

There is also other life hidden in the dry valleys. High in the mountains, above the valley floor, are lichens living within the pore spaces of sandstone rocks (see Color Plate 6.8). The rocks are translucent and dark. Light is absorbed by the rock surface and some light penetrates inside, which allows the lichen to photosynthesize. On a sunny summer day, the rocks can be as much as 15 degrees warmer than the air; snow on the rocks melts and percolates into the pore spaces. Inside this warm, wet environment the cryptoendolithic lichens thrive while outside conditions are intolerably dry and cold (Friedmann, 1982).

The life in the Antarctic dry valleys shows that microbial life can persist well after the average temperature has fallen below freezing. Using these systems as models, we can estimate how long life may have persisted on Mars. The key factor in maintaining an ice-covered lake habitat would be to have the daily average temperatures at the peak of the summer above freezing. On Mars today, the summertime noon temperatures may rise above freezing, but the average daily temperatures remain well below freezing. With a thicker atmosphere, the extremes are averaged out.

With a climate model, we can calculate that peak summer temperatures above freezing would occur on Mars for yearly averaged temperatures as low as −35°C. The warmest summers would occur in the hemisphere that had summer when Mars was at perihelion, the closest point to the Sun in Mars' **eccentric** orbit. Climate models suggest that about a half-billion years would be needed for the temperature on Mars to drop from 0°C to −35°C (McKay and Davis, 1991). Life on Mars could have survived in ice-covered lakes such as those we find in Antarctica long after the rest of the Martian surface had become lifeless. Lichens growing in sandstone-like rocks — if such exist on Mars — could survive until the yearly averaged temperature fell by another 10 degrees.

On Mars there are several locations in which possible ancient lakes have been identified. One of the most interesting sites is Hebes Chasma — a box canyon, located on the Martian equator, that is part of the Valles Marineris canyon system (see Color Plate 6.9). Within the canyon is a plateau composed of clearly discernable layers. The morphology of the layers and the closed nature of the canyon indicates that the deposits may be carbonate precipitated in a lake that filled the canyon billions of years ago. An ice-covered lake in this canyon may have been the last refuge for life on a dying world.

LIFE IN THE SOLAR SYSTEM

Other than Earth and early Mars, there is no evidence of liquid water on the surface of any other world in our solar system. However, there have been speculations regarding liquid water on Europa and on comets. On Europa, one of the big Galilean moons of Jupiter, there is the possibility that an ocean of liquid water is present beneath a relatively thin ice cover (Reynolds et al., 1983, 1987). However, unlike on Mars, there is no direct evidence for its existence. If there is liquid water on Europa, the ice shell that overlies it would be expected to crack occasionally and allow sunlight to reach the water column. Perhaps life could survive in a manner similar to that of **marine diatoms** that grow under the Antarctic sea ice (Reynolds et al., 1983).

Comets are a class of objects that also may have had liquid water at one time (Irvine et al., 1980; Wallis, 1980). It is possible that **radiogenic heating** would have produced liquid water cores in large comets for a brief period after their formation (Prialnik et al., 1987). Other than this possible brief occurrence of a liquid state, water in comets is expected to be in the form of ice. Ice on Earth is devoid of life except where some liquid water is present, which suggests that it is unlikely that life can grow on cometary ice. However, comets may harbor dormant life (McKay, 1996).

Titan is the largest moon of the planet Saturn and the only moon with an atmosphere. Although Titan has copious organic material in its atmosphere, it lacks the essential condition for life, liquid water. Its surface temperature is too low for water to be liquid. Although Titan may have seas of methane and ethane, these substances do not readily dissolve organic material. Thus, prospects for life are remote.

The possibility of life in the water clouds of the atmospheres of the giant planets, such as Jupiter, is not supported by observations of clouds on Earth. Earth clouds are probably many times more habitable than clouds on Jupiter and Saturn, yet clouds on Earth are not sites of microbial growth. Speculations of floaters, drifters, and other exotic cloud-based life-forms in the atmosphere of Jupiter (Sagan and Salpeter, 1976) seem to be unsupported.

Other aspects of the stellar system in which a planet forms may be crucial to the origin of life. These aspects include the source of organic material as well as the source of water, carbon dioxide, and other necessary compounds. In the standard theory for the origin of life, organic molecules are first produced by nonbiological sources. Production requires that the planet form in such a way that the early atmosphere is rich in reduced gases such as methane (CH_4) and ammonia (NH_3), which would allow for production of organics in the early atmosphere. Alternatively, organics can be deposited on a planet by organic-rich objects from the outer reaches of the stellar system. The interstellar clouds from which a star and its planetary system form are known to be rich in organics. In our solar system, objects beyond 3 **astronomical units** (AU) from the Sun appear to retain much of their organic heritage. Objects closer to the Sun are depleted in carbon. Figure 6.10 shows the distribution of carbon in our solar system.

Asteroids and comets represent a storehouse of organics that were available for the origin of life (Chyba and Sagan, 1992). For this process to have been duplicated around another star, several conditions must have been satisfied. First, the temperature and shock waves in the collapsing nebula from which the star and planets formed had to be consistent with the preservation of interstellar organic material among the outer planets. Second, small mobile bodies such as asteroids and comets that can transport this material to the inner solar system had to exist. Finally, giant planets the size of Neptune or larger had to exist to scatter the small bodies into the realm of the inner planets. At present we do not know whether these conditions are common or rare around other stars, but the presence of life may depend on them.

The evolution of advanced life may require even more exacting conditions and could be closely tied to the planetary neighborhood. One main hazard appears to be environmentally damaging impacts. The frequency of such catastrophes may need to be low enough to allow life to develop, but perhaps frequent enough to promote occasional rearrangements in the pattern of evolution. Indeed, the extinction of the dinosaurs 65 million years ago, which is strongly tied to a comet impact, may have been crucial in the development of mammals, humans, and, hence, intelligence. Impacts large enough to cause mass extinctions at too frequent a rate may stymie the development of advanced forms, but too low a rate may result in evolutionary stasis that leads nowhere. The rate of impacts is controlled by the concentration of impacting objects. In our solar system, the presence of the giant Jupiter resulted in the gravitational scattering and, ultimately, the removal of virtually all small bodies left after the formation of the planets. In a planetary system without a Jupiter-sized object, the impact flux into the inner solar system could be 1000 times larger than that in our solar system (Wetherill, 1994). Interestingly, preliminary studies show that Jupiter-sized objects may be rare (Zuckerman et al., 1995; Walker et al., 1995).

Climate stability is another factor in the development of advanced life. Certain levels of climate change may promote evolution and force the development of coping strategies that lead ultimately to intelligence. Too much variability, however, may destroy advanced life-forms. An important control on climate is **obliquity**, the tilt of a planet's spin axis with respect to its orbit around a star. Earth's substantial obliquity (~23°) causes the seasons. For Earth, changes in obliquity are moderated by the presence of the Moon. Without a large close moon, huge obliquity changes might occur; obliquity would vary chaotically (Laskar et al., 1993). Although the deleterious effect that large impacts have on advanced life was clearly demonstrated 65 million years ago, the effect of huge obliquity changes on advanced life is not clear.

Impacts and climate change affect only advanced life, which is typically large and relatively sensitive to ecological changes. Microorganisms are at no risk from these changes. To alter the microbial communities of Earth, an impact would have to be energetic enough to heat the entire planet to sterilizing temperatures (Maher and Stevenson, 1988; Sleep et al., 1989). On Earth, the most recent such impact probably occurred more than 3.8 billion years ago (Sleep et al., 1989) but could occur even later in a planetary

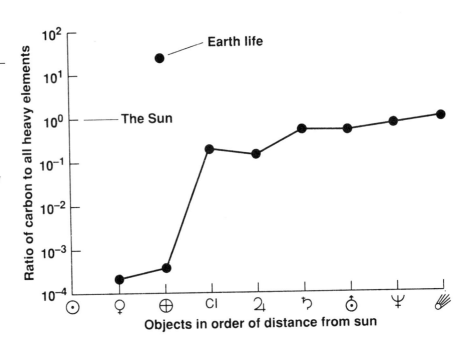

FIGURE 6.10

Carbon Abundance in the Solar System: Ratio of carbon atoms to total heavy atoms (heavier than He) for various solar system objects, which illustrates the depletion of carbon in the inner solar system. The x-axis is not a true distance scale, but the objects are ordered by increasing distance from the sun. Mars is not shown because the size of its carbon reservoir is unknown. C1 refers to most common class of asteroids. (From McKay, 1991.)

system without a large planet such as Jupiter. Similarly, cold climates would have to be as severe as those on Mars and persist for many tens of millions of years for microbial life to be eradicated. Advanced life is much easier to destroy than microbial life. (The bigger they are, the harder they fall.)

SEARCHING FOR LIFE AROUND OTHER STARS

If life-bearing, habitable worlds exist around other stars, how could we detect them? On Earth, there have been three distinct stages of progression to advanced life: 1) microbial anaerobes, 2) oxygen in the atmosphere, and 3) intelligence and technology (radio transmissions). These stages are shown in Figure 6.11 along with methods for detecting signs of life during each stage. If a stage of advanced life is characterized by broadcasting radio waves, detection may be fairly easy. This is the principle behind the Search for Extraterrestrial Intelligence (SETI) program. The detection of intelligent radio emissions completely obviates the search for microbial life, liquid water, and planets within another star's habitable zone. The search is replaced with a dialogue. Unfortunately, the duration of technological civilization on Earth has been infinitesimal compared with that of life overall. Furthermore, intelligence may be rare (Hart, 1975; McKay, 1995), which suggests that other search strategies may need to be conducted in parallel with SETI.

The presence of oxygen in the atmosphere of an extrasolar planet can be detected most readily by searching for the spectral signature of ozone as the planet transits its central star (Owen, 1980; Schneider, 1994). Taking Earth as an example, a search for oxygen or ozone is a strategy that would have yielded good results over less than half Earth's biotic history. To achieve a more inclusive result, one must search for the presence of liquid water (which on Earth has coincided with biology). A search for liquid water must have two components. First, water vapor in the atmosphere must be present. This, however, is not enough, because a search of our solar system would show water in the atmospheres of

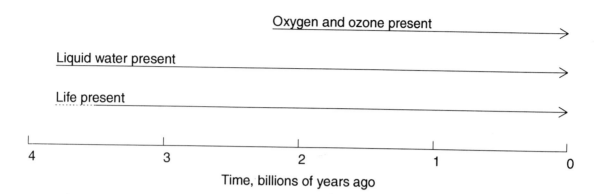

FIGURE 6.11

Timeline showing methods for detecting life and habitable conditions on Earth from a nearby star. Liquid water has always been an indicator of habitability on Earth. Oxygen has been present for about half Earth's biological history; radio emissions have been present for only the past 75 years. A similar approach could be used from Earth to detect life and liquid water on extrasolar planets.

Venus, Earth, and Mars. Detection of atmospheric water vapor would have to be complemented by the measurement of the planetary surface temperature. This could be done, despite the presence of obscuring haze and clouds, by using microwave radiation to penetrate the atmosphere. The intensity of the microwave emission depends only on the planet's surface temperature. For Venus, Earth, and Mars, the results would be 440°C, 15°C, and –60°C, respectively. The sole habitability of Earth would be established.

IT'S LIFE, JIM, BUT NOT AS WE KNOW IT

Throughout this chapter we have considered life only as we find it on Earth: water-based, carbon-built life. We have assumed that the essential features of life would be the same on another planet on which the main energy source was sunlight, the primary liquid was water, and the most naturally reactive element was carbon, found primarily in the form of CO_2. We have only one example of life; thus, it is certainly premature to deduce that any characteristics of this life are absolute requirements for life elsewhere. On the other hand, a fruitful search strategy must be based on some model of life. Because we have only one example, the safest strategy is to search in similar environments for life of a similar type. This is a purely practical consideration, and does not imply that life very different from Earth-life cannot exist. (See Color Plate 6.12.)

Two main suggestions for alternative biochemical systems are the substitution of ammonia for water as the solvent for life's molecules, and the substitution of silicon for carbon as the basic structural element. Neither suggestion can be disproved at the present time, but neither can form the basis for a search strategy that is as well developed as the search for water-carbon life. We must first seek our cousins, and from there a broader understanding that can allow us to see strangers.

REFERENCES

Biemann, K., Oro, J., Toulmin, P., III, et al. 1977. The search for organic substances and inorganic volatile compounds in the surface of Mars. *Journal of Geophysical Research 82*: 4641–4658.

Cairns-Smith, A. G. 1982. *Genetic Takeover and the Mineral Origins of Life* (Cambridge, UK: Cambridge University Press).

Chyba, C., and Sagan, C. 1992. Endogenous production, exogenous delivery and impact-shock synthesis of organic molecules: An inventory for the origins of life. *Nature 355*: 125–132.

Clow, G. D., McKay, C. P., Simmons, G. M., Jr., and Wharton, R. A., Jr. 1988. Climatological observations and predicted sublimation rates at Lake Hoare, Antarctica. *Journal of Climate 1*: 715–728.

Davis, W. L., and McKay, C. P. 1996. Theories for the origin of life and applications to Mars. *Origins of Life in Evolutionary Biosphere* (in press).

DesMarais, D. J., Strauss, H., Summons, R. E., and Hayes, J. M. 1992. Carbon isotope evidence for the stepwise oxidation of the Proterozoic environment. *Nature 359*: 605–609.

Eigen, M., Lindemann, B. F., Tietze, M., Winkler-Oswatitsch, R., Dress, A., and von Haeseler, A. 1989. How old is the genetic code? Statistical geometry of tRNA provides an answer. *Science 244*: 673–679.

Encrenaz, T. 1984. Primordial matter in the outer solar system: A study of its chemical composition from remote spectroscopic analysis. *Space Science Review 38*: 35–87.

Friedmann, E. I. 1982. Endolithic microorganisms in the Antarctic cold desert. *Science 215*: 1045–1053.

Hart, M. H. 1975. An explanation for the absence of extraterrestrials on Earth. *Quarterly Journal of the Royal Astronomical Society 16*: 128–135.

Irvine, W. M., and Knacke, R. F. 1989. The chemistry of interstellar gas and grains. In S. K. Atreya, J. B. Pollack, and M. S. Matthews (Eds.), *Origin and Evolution of Planetary and Satellite Atmospheres* (Tucson: University of Arizona Press), pp. 3–34.

Irvine, W. M., Leschine, S. B., and Schloerb, F. P. 1980. Thermal history, chemical composition and relationship of comets to the origin of life. *Nature 283*: 748–749.

Joyce, G. 1989. RNA evolution and the origins of life. *Nature 338*: 217–224.

Kissel, J., and Krueger, F. R. 1987. The organic component in the dust from comet Halley as measured by the PUMA mass spectrometer on board Vega 1. *Nature 326*: 755–760.

Knoll, A. H. 1985. The Precambrian evolution of terrestrial life. In: M. D. Papagiannis (Ed.), *The Search for Extraterrestrial Life: Recent Developments* (IAU Symposium #112) (Dordrecht: Reidel), pp. 201–211.

Kushner, D. 1981. Extreme environments: Are there any limits to life? In: C. Ponnamperuma (Ed.), *Comets and the Origin of Life* (Dordrecht: D. Reidel), pp. 241–248.

Laskar, J., Joutel, F., and Robutel, P. 1993. The stabilization of the Earth's obliquity by the Moon. *Nature 361*: 615–617.

Lazcano, A., and Miller, S. L. 1994. How long did it take for life to begin and evolve to cyanobacteria? *Journal of Molecular Evolution 39*: 546–554.

Lehninger, A. L. 1975. *Biochemistry* (New York: Worth).

Maher, K. A., and Stevenson, D. J. 1988. Impact frustration of the origin of life. *Nature 331*: 612–614.

McKay, C. P. 1991. Urey Prize paper: Planetary evolution and the origin of life. *Icarus 91*: 93–100.

McKay, C. P. 1993. Relevance of Antarctic microbial ecosystems to exobiology. In: E. I. Friedmann (Ed.), *Antarctic Microbiology* (New York: Wiley-Liss), pp. 593–601.

McKay, C. P. 1995. Time for intelligence on other planets. In: L. R. Doyle (Ed.), *Circumstellar Habitable Zones: Proceedings of the First International Conference* (Menlo Park: Travis House) (in press).

McKay, C. P. 1996. Life in comets. In: P. Thomas, C. P. McKay, and C. F. Chyba (Eds.), *Comets and the Origin and Evolution of Life* (Berlin: Springer-Verlag; in press).

McKay, C. P., and Stoker, C. R. 1989. The early environment and its evolution on Mars: Implications for life. *Review of Geophysics 27*: 189–214.

McKay, C. P., and Davis, W. L. 1991. Duration of liquid water habitats on early Mars. *Icarus 90*: 214–221.

McKay, C. P., Clow, G. A., Wharton, R. A., Jr., and Squyres, S. W. 1985. Thickness of ice on perennially frozen lakes. *Nature 313*: 561–562.

McKay, C. P., Clow, G. D., Andersen, D. T., and Wharton, R. A., Jr. 1994. Light transmission and reflection in perennially ice-covered Lake Hoare, Antarctica. *Journal of Geophysical Research 99*: 20427–20444.

Melosh, H. J. 1988. The rocky road to paspermia. *Nature 332*: 687–688.

Miller, S. L. 1992. The prebiotic synthesis of organic compounds as a step toward the origin of life. In: J. W. Schopf (Ed.), *Major Events in the History of Life* (Boston: Jones and Bartlett), pp. 1–28.

Oró, J. 1961. Comets and the formation of biochemical compounds on the primitive Earth. *Nature 190*: 389–390.

Owen, T. 1980. The search for early forms of life in other planetary systems: Future possibilities afforded by spectroscopic techniques. In: M. D. Papagiannis (Ed.), *Strategies for the Search for Life in the Universe* (Boston: Reidel), pp. 177–185.

Prialnik, D., Bar-Nun, A., and Podolak, M. 1987. Radiogenic heating of comets by ^{26}Al and implications for their time of formation. *Astrophysical Journal 319*: 993–1002.

Reynolds, R. T., McKay, C. P., and Kasting, J. F. 1987. Europa, tidally heated oceans, and habitable zones around giant planets. *Advances in Space Research 7*(5): 125–132.

Reynolds, R. T., Squyres, S. W., Colburn, D. S., and McKay, C. P. 1983. On the habitability of Europa. *Icarus 56*: 246–254.

Rivera, M. C., and Lake, J. A. 1992. Evidence that eukaryotes and eocyte prokaryotes are immediate relatives. *Science 257*: 74–76.

Russell, D. A., and Séguin, R. 1982. Reconstructions of the small Cretaceous Theropod *Stenonychosaurus inequalis* and a hypothetical dinosauroid. In: *Syllogeus #37* (Ottawa: National Museum of Natural Sciences), 43 pages.

Sagan, C., and Salpeter, E. E. 1976. Particles, environments, and possible ecologies in the Jovian atmosphere. *Astrophysical Journal Supplement Series 32*: 737–755.

Sagan, C., Khare, B. N., and Lewis, J. S. 1984. Organic matter in the Saturn system. In: T. Gehrels and M. S. Matthews (Eds.), *Saturn* (Tucson: University of Arizona Press), pp. 788–807.

Schneider, J. 1994. On the search for O_2 in the atmosphere of extrasolar planets. *Astrophysical Space Science 212*: 321–325.

Schidlowski, M. 1988. A 3800-million-year isotopic record of life from carbon in sedimentary rocks. *Nature 333*: 313–318.

Schopf, J. W. 1993. Microfossils of the early Archean apex chert: New evidence for the antiquity of life. *Science 260*: 640–646.

Sleep, N. H., Zahnle, K. J., Kasting, J. F., and Morowitz, H. J. 1989. Annihilation of ecosystems by large asteroid impacts on the early Earth. *Nature 342*: 139–142.

Walker, G. A. H., Walker, A. R., Irwin, A. W., et al. 1995. A search for Jupiter — Mass companions to nearby stars. *Icarus 116*: 359–375.

Wallis, M. K. 1980. Radiogenic melting of primordial comet interiors. *Nature 284*: 431–433.

Wetherill, G. W. 1994. Possible consequences of absence of "Jupiters" in planetary systems. *Astrophysical Space Science 212*: 23–32.

Woese, C. R. 1987. Bacterial evolution. *Microbiological Reviews 51*: 221–271.

Zent, A. P., and McKay, C. P. 1994. The chemical reactivity of the martian soil and implications for future missions. *Icarus 108*: 146–157.

Zuckerman, B., Forveille, T., and Kastner, J. H. 1995. Inhibition of giant-planet formation by rapid gas depletion around young stars. *Nature 373*: 494–496.

CHAPTER 7

FUTURE OF THE UNIVERSE

■

Andrei Linde

BIG BANG THEORY VERSUS INFLATIONARY COSMOLOGY

For many years, cosmologists believed that the Universe from the very beginning looked like an expanding ball of fire. This explosive beginning of the Universe was called the **Big Bang**. The standard Big Bang theory asserts that the Universe was born at some moment $t = 0$ about 15 billion years ago, in a state of infinitely high temperature and **density** (a *cosmological singularity*). Of course, we cannot really speak about infinite temperature and density. It is usually assumed that the standard description of the Universe in terms of space and time becomes possible when density drops below the so-called Planck density, approximately 10^{94} g/cm^3. The temperature of the expanding Universe gradually decreased as a reciprocal of its size and finally evolved into the relatively cold Universe in which we now live.

About 15 years ago, a different scenario of the evolution of the Universe was proposed. The main idea of the new scenario was that the Universe at the very early stages of its evolution came through a stage of inflation, an exponentially rapid expansion in an unstable, vacuum-like state. This stage could have been very short, but the Universe within this time grew exponentially in time. At the end of inflation, the energy of the vacuum-like state transformed into thermal energy, the Universe became hot, and its subsequent evolution could be described by the standard hot Big Bang theory.

Originally, the **inflationary universe scenario** looked like an interesting piece of science fiction. Now, however, many scientists believe that inflationary cosmology is the only internally consistent cosmological theory that we have. Fifteen years of development have made the basic principles of this theory much simpler than they were initially; some of the consequences of this theory have become even more surprising. Its investigation has gradually changed the whole cosmological paradigm, including the very notion of the Big Bang and our basic ideas about the global structure of the Universe. A popular account of inflationary cosmology accompanied by the results of computer simulations can be found in Linde (1994); for a detailed review of the theory, see Linde (1990a, 1990b). Inflationary theory radically changed our point of view not only on the origin of our Universe but also on its future evolution.

BIG BANG THEORY AND THE FUTURE OF THE UNIVERSE

There have been several important stages in the development of twentieth century cosmology. The first began in the 1920s, when Alexander Friedmann from St. Petersburg solved Einstein's equations of the general theory of relativity describing a homogeneous universe. Friedmann obtained three different classes of solutions, corresponding to open, flat, and closed universes. To understand the physical meaning of these solutions we can use a simple Newtonian analogy.

Let us consider a spherically symmetric cloud of particles moving away from the center of the sphere. If the density of the gas of particles is very small, their gravitational attraction is insignificant and the cloud of particles continues expanding indefinitely. On the other hand, if the density of particles is very high, gravitational forces between them are strong and all the particles eventually fall back down toward the center of the sphere. There is also an intermediate regime, on the boundary between these two regimes. If the density of particles is equal to some **critical density**, particles continue moving away from the center and their speed gradually vanishes, but it reaches zero only infinitely far into the future.

Something similar happens in the expanding Universe. If the density of particles is sufficiently small, expansion of the Universe continues forever (open Universe). The second case (a dense cloud of particles) is similar to the model of a closed Universe. In the beginning, a closed Universe expands, then, at some moment, the size of the Universe reaches its maximum and the Universe collapses. A flat Universe has a critical density, which is just enough to gradually slow down expansion of the Universe, but is not enough to lead to its collapse.

This theory was extremely successful in explaining various features of our world. It became a standard paradigm of modern cosmology when Penzias and Wilson, in 1965, discovered microwave background radiation with a temperature of 2.7 kelvins (K). The radiation was interpreted as a remnant of the primordial cosmic fire that still surrounds us in the form of dim light coming from all points on the sky (see Chapter 1).

Among many predictions of the Big Bang theory, one was rather disturbing. If the Universe is closed, at some moment it should collapse. In the process of collapse, the microwave background radiation should be compressed, its temperature should grow, and finally the whole Universe, together with all its stars and planets, should disappear in the same cosmic fire from which it originated. This is a very uncomfortable picture, so we might feel some temporary relief thinking about the possibility that our Universe is not closed, but is flat or open. In this case, the Universe continues expanding forever. However, in the course of time, galaxies move far away from each other, stars stop burning, and all inhabitants of an open Universe die in the solitude of cold empty space.

We could hope that life may survive for a longer time in the vicinity of huge evaporating **black holes**, which would supply us with energy until they disappear. Perhaps a more promising idea would be to take energy from rotating black holes. But even this energy is not unlimited.

The final blow to our optimism was made by the theory of proton decay. According to the modern theory of elementary particles, protons and neutrons, which constitute the main bulk of matter in our bodies, should eventually decay into such particles as positrons (antielectrons) and neutrinos. This process may take as long as 10^{35} years, but when it happens, nothing can save us from disappearing unless we find a way to transplant our souls to the clouds of **electrons**, **positrons**, **photons**, and **neutrinos** populating the cold Universe devoid of stars and planets.

PROBLEMS OF THE BIG BANG THEORY

The Big Bang theory did not allow much room for optimism concerning our future. However, 15 years ago, physicists realized that the theory should be modified because it was plagued by many complicated problems related not only to our future, but to our past and present as well. For example, the standard Big Bang theory combined with the modern theory of elementary particles predicts the existence of many superheavy stable particles carrying magnetic charge: magnetic monopoles. These objects have a typical mass 10^{16} times that of the proton. According to the standard Big Bang theory, monopoles should appear at the very early stages of the evolution of the Universe, and they should now be as abundant as protons. In that case, the mean density of matter in the Universe would be about 15 orders of magnitude higher than its present value of about 10^{-29} g/cm^3.

Originally, we hoped that this problem would disappear when more complicated theories of elementary particles were considered. Unfortunately, despite the rapid development of elementary particle physics, the monopole problem remained unsolved, and many new ones have been added: the gravitino problem, Polonyi field problem, domain wall problem, axion window problem, and so on. Physicists were forced to look more attentively at the basic assumptions of the standard cosmological theory. We have found that many of the assumptions are very suspicious.

The main problem of the Big Bang cosmology is the very existence of the Big Bang. What was before the Big Bang? Where did the Universe come from? If space-time did not exist for times less than 0, how could everything appear from nothing? What appeared first: the Universe or the laws determining its evolution? When we were born, the laws determining our development were written in the genetic code of our parents. But where were the laws of physics written when there was no Universe?

The problem of cosmological singularity still remains the most difficult problem of modern cosmology. However, we can now look at it from a totally different point of view. At school we are taught that two parallel lines never cross. However, general relativity tells us that our Universe is curved. The Universe may be open, in which case parallel lines diverge from one another, or it may be closed, and parallel lines cross each other like meridian lines on a globe. The only natural length parameter in general relativity is the Planck length l_p approximately 10^{-33} cm. At smaller distances, the standard concept of space becomes inapplicable because of large **quantum fluctuations**. Therefore, we would expect our space to be very curved, with a typical radius of curvature about 10^{-33} cm. We see, however, that our Universe is just about flat on a scale of 10^{28} cm, the radius of the observable part of the Universe. The results of our observations differ from our theoretical expectations by more than 60 orders of magnitude!

Why are there so many different people on Earth? Well, Earth is large, so it can accommodate a lot of people. But why is Earth so large? (In fact, it is extremely small compared with the whole Universe.) Why is the Universe so large? A typical answer is that the Universe is large because it is a Universe, so it should be large, should it not? However, let us consider the Universe of a typical size l_P just emerging from the Big Bang. Let us assume that this Universe had the greatest possible density at which we can still describe it in terms of usual space and time. This is the so-called Planck density, 10^{94} g/cm^3, which corresponds to matter consisting of particles displaced at the Planck distance l_P from each other. We do not know how to describe matter with a greater density. But if we take the Planck size universe with the Planck density, we can easily understand, by definition of the Planck density, that such a universe can contain only a few particles. The standard assumption of the Big Bang cosmology is that the total number of elementary particles in the Universe almost does not change during the expansion of the Universe. Thus, a typical universe should contain one particle, maybe ten particles, but not 10^{88} particles, which is the number contained in the part of the Universe we see now. This is a contradiction by 88 orders of magnitude.

The standard assumption of the Big Bang theory is that all parts of the Universe began their expansion simultaneously, at the moment $t = 0$. But how could different parts of the Universe synchronize the beginning of their expansion if they did not have any time for it? Who gave the command?

Our Universe, on a very large scale, is extremely homogeneous. On a scale of 10^{10} **light-years**, the distribution of matter departs from perfect homogeneity by less than 1 part in 100,000. For a long time nobody had any idea why the Universe was so homogeneous. But those who do not have good ideas sometimes have good principles. One of the cornerstones of the standard cosmology was the *cosmological principle*, which asserts that the Universe must be homogeneous. However, the Universe contains stars, galaxies, and other important deviations from homogeneity. We have two opposite problems to solve. First, we must explain why our Universe is so homogeneous, and then we should suggest some mechanism that produced galaxies.

All these problems (and others) are extremely difficult. That is why it is very encouraging that most of these problems can be resolved in the context of one simple theory of the evolution of the Universe — the inflationary scenario. Meanwhile, some of its consequences are even more surprising. Instead of the Universe looking like a single expanding ball created in the Big Bang, we envisage it now as a huge, growing **fractal** consisting of many inflating balls that are producing new balls that are producing new balls, ad infinitum.

INFLATIONARY COSMOLOGY

Unified Theories of Elementary Particles

To explain basic features of inflationary cosmology, we make an excursion into the theory of elementary particles. Rapid progress of this theory during the past 2 decades became possible after physicists found a way to unify **weak nuclear**, **strong nuclear**, and electromagnetic interactions. Electrically charged particles interact with each other by creating an electromagnetic field around them. Small excitations of this field are called photons. Photons do not have any mass, which is the main reason electrically charged particles easily interact with each other at a very large distance. Scientists believe that weak and strong interactions are mediated by similar particles. For example, weak interactions are mediated by experimentally discovered particles called W and Z. However, whereas photons are massless particles, the particles W and Z are extremely heavy; it is very difficult to produce them. That is why weak interactions are so weak. To obtain a unified description of weak and electromagnetic interactions despite the obvious difference in properties of photons and the W and Z particles, physicists introduced scalar fields ϕ, which will play the central role in our discussion.

The theory of scalar fields is simple. The closest analog of a scalar field is the electrostatic potential Φ. Electric and magnetic fields E and H appear only if this potential is inhomogeneous or if it changes in time. If the whole Universe had the same electrostatic potential, say, 110 volts, nobody would notice; it would be just another vacuum state. Similarly, a constant scalar field ϕ looks like a vacuum state; we do not see it even if we are surrounded by it.

The main difference is that the constant electrostatic field Φ does not have its own energy, whereas the scalar field ϕ may have potential energy density $V(\phi)$. If $V(\phi)$ has one minimum at $\phi = \phi_0$, the whole Universe eventually becomes filled by the field ϕ_0. This field is invisible, but if it interacts with particles W and Z, they become heavy. Meanwhile, if photons do not interact with the scalar field, they remain light. Therefore, we begin with a theory in which all particles initially are light, and there is no fundamental difference between weak and electromagnetic interactions. The difference appears later, when the Universe becomes filled by the scalar field ϕ. At this moment, the symmetry between different types of fundamental interactions becomes broken. This is the basic idea of all unified theories of weak, strong, and electromagnetic interactions.

If the potential energy density $V(\phi)$ has more than one minimum, the field ϕ may occupy any of them. This means that the same theory may have different vacuum states, corresponding to different types of symmetry breaking between fundamental interactions, and, as a result, to different laws of physics of elementary particles. To be more accurate we should speak about different laws of low-energy physics. At an extremely high energy, the difference in masses becomes not very important, and the initial symmetry of all fundamental interactions reveals itself again.

In many theories of elementary particles now popular, it is assumed that space-time originally had considerably more than four dimensions, but extra dimensions have been compactified, shrunk to a very small size. That is why we cannot move in the corresponding directions and our space-time looks four-dimensional. However, why did compactification stop with four effective space-time dimensions, not two or five? Moreover, in the higher-dimensional theories, compactification may occur in a thousand different ways. According to these theories, the values of particle masses and the relative strengths of the fundamental forces strongly depend on the way compactification occurs. Constructing such theories that admit only one type of compactification and only one way of symmetry breaking became increasingly difficult.

This adds to our list of problems yet another problem — the uniqueness problem. Its essence was formulated by Albert Einstein: "What really interests me is whether God had any choice in the creation of the world." A few years ago it would have seemed rather meaningless to ask why our space-time is four-dimensional, why the gravitational constant is so small, why the proton is 2000 times heavier than the electron, and so forth. Now these questions have acquired a simple physical meaning, and we can no longer ignore them. Inflationary theory may help us answer these questions.

The Simplest Version of Inflationary Theory — Chaotic Inflation

According to the Big Bang theory, the rate of expansion of the Universe given by the Hubble constant $H(t)$ is approximately proportional to the square root of its density (see Chapter 1). If the Universe is filled with ordinary matter, its density rapidly decreases as the Universe expands. Therefore, the expansion of the Universe rapidly slows as its density decreases. The rapid decrease of the rate of expansion is the main reason for our problems with the standard Big Bang theory. However, because of the equivalence of mass and energy established by Einstein ($E = mc^2$), the potential energy density $V(\phi)$ of the scalar field ϕ also contributes to the rate of expansion of the Universe. In certain cases, the energy density $V(\phi)$ decreases much more slowly than the density of ordinary matter, which may lead to a stage of extremely rapid expansion (inflation) of the Universe.

Let us consider the simplest model of a scalar field ϕ with a mass m and with the potential energy density $V(\phi) = \frac{m^2}{2}\phi^2$. Because this function has a minimum at $\phi = 0$, we expect that the scalar field ϕ will oscillate near the minimum. This is indeed the case if the Universe does not expand. However, we can show that in a rapidly expanding Universe an equation describing the motion of the scalar field acquires an additional term. The additional term implies that the scalar field moves down very slowly, as a ball in a viscous liquid, viscosity being proportional to the speed of expansion (see Color Plate 7.1).

Now we have only one step to understand where inflation comes from. If the scalar field ϕ initially was large, its energy density $V(\phi)$ was also large, and the Universe expanded very rapidly. Because of rapid expansion, the scalar field was moving to the minimum of $V(\phi)$ very slowly. Therefore, at this stage, the energy density $V(\phi)$, unlike the density of ordinary matter, remained almost constant, and expansion of the Universe continued with much greater speed than that in the old cosmological theory — the size of the Universe in this regime grows approximately as e^{Ht}, where H is the Hubble "constant" proportional to the square root of $V(\phi)$.

For those who want to understand this effect at a more formal level, these are the two equations that describe an inflationary universe: $\ddot{\phi} + 3H\dot{\phi} = -dV/d\phi$, and $H^2 = \frac{8\pi}{3M_P^2}V(\phi)$. The second equation is a slightly simplified Einstein equation for the scale factor (radius) of the universe $a(t)$; H is a Hubble constant, $H = \dot{a}/a$ (see Chapter 1). If $V(\phi)$ is approximately constant during a sufficiently long period of time, the second equation has a solution $a(t) \sim e^{Ht}$, which describes an exponentially rapid expansion (inflation) of the Universe.

The stage of self-sustained, exponentially rapid, expansion of the Universe was not very long. In a realistic version of our model, its duration could be as short as 10^{-35} seconds. When the energy density of the field ϕ becomes sufficiently small, viscosity becomes small, inflation ends, and the scalar field ϕ begins to oscillate near the minimum of $V(\phi)$. A rapidly oscillating classical field loses its energy by creating pairs of elementary particles. The particles interact with each other and come to a state of thermal equilibrium with some temperature T. From this time on, the Universe (or, to be more precise, the part of the Universe where inflation ends) can be described by the standard Big Bang theory (see Figure 7.2).

The main difference between inflationary theory and the old cosmology becomes clear when we calculate the size of a typical inflationary domain at the end of inflation. Investigation shows that even if the size of the part of the inflationary universe at the beginning of inflation in our model was as small as l_P (10^{-33} cm) after 10^{-35} seconds of

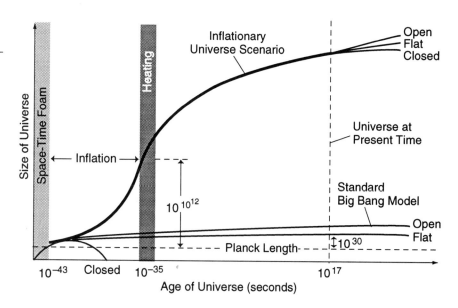

FIGURE 7.2

Evolution of the Universe differs in the chaotic inflation scenario and the standard Big Bang theory. Inflation increases the size of the Universe by $10^{10^{12}}$, so that even parts as small as 10^{-33} cm (the Planck length) exceed the radius of the observable Universe, 10^{28} cm. Inflation also predicts space to be mostly flat, in which parallel lines remain parallel. (Parallel lines in a closed universe intersect; in an open one, they ultimately diverge.) In contrast, the original, hot Big Bang explosion would have increased a Planck-size Universe to only 0.001 cm and would have led to different predictions about the geometry of space.

inflation, the domain acquires a huge size of approximately $10^{10^{12}}$ cm! These numbers are model-dependent, but in all realistic models this size appears to be many orders of magnitude greater than the size of the part of the Universe that we see now, approximately 10^{28} cm. Inflationary theory immediately solves most problems of the old cosmological theory.

Our Universe is homogeneous because all inhomogeneities were stretched $10^{10^{12}}$ times. The density of primordial monopoles, domain walls, gravitinos, and other undesirable defects becomes exponentially diluted by inflation. The Universe becomes enormously large. Even if it were a closed Universe of a size about 10^{-33} cm, after inflation the distance between its South and North poles becomes many orders of magnitude greater than 10^{28} cm. We see only a tiny part of the huge cosmic balloon. That is why the Universe looks so flat. That is why nobody has ever seen how parallel lines cross. That is why we do not need expansion of the Universe to begin simultaneously in 10^{88} different causally disconnected domains of a Planck size. One such domain is enough to produce everything that we see now.

Brief History of Inflationary Cosmology

Inflationary theory did not always look so conceptually simple. Attempts to obtain the stage of exponential expansion of the Universe have a long history. Unfortunately, because of artificial barriers for exchange of information, which existed earlier in the Soviet Union, the history is only partially known to American readers.

The first realistic version of inflationary theory was proposed in 1979 by Alexei Starobinsky of the Landau Institute, Moscow (Starobinsky, 1979, 1980). Starobinsky's model created a sensation among Russian astrophysicists, and for 2 years remained the main topic of discussion at all conferences on cosmology in the Soviet Union. However, Starobinsky's model was rather complicated, having been based on the theory of anomalies in quantum gravity. In 1981, Alan Guth of the Massachusetts Institute of Technology (MIT) suggested that the hot Universe at some intermediate stage could expand exponentially, because it was in an unstable, supercooled state (Guth, 1981). Guth's model was based on the theory of cosmological **phase transitions** with supercooling, which was developed in the mid-1970s by Kirzhnits and Linde at the Lebedev Institute, Moscow (Kirzhnits and Linde, 1972, 1976). This scenario was attractive, and it had a very clear physical motivation. Unfortunately, as Guth found, the Universe after inflation in his scenario becomes extremely inhomogeneous. After a year of investigation of his model, Guth finally renounced it in his paper with Eric Weinberg of Columbia University (Guth and Weinberg, 1983). In 1982 Linde introduced the so-called new inflationary universe scenario (Linde, 1982), which later was also discovered by Andreas Albrecht and Paul Steinhardt at the University of Pennsylvania (Albrecht and Steinhardt, 1982). This scenario was free of the main problems of the model suggested by Guth, but it was still rather complicated; no realistic versions of this scenario have thus far been proposed.

A year later Linde realized that inflation can naturally occur in many theories of elementary particles, including the simplest theory of the scalar field. There is no need for quantum gravity effects, phase transitions, and supercooling. We should consider all possible chaotic distributions of the scalar field in the early Universe, and then check whether some lead to inflation. Places in which inflation does not occur remain small, but the domains in which inflation takes place become exponentially large and give the main contribution to the total volume of the Universe. This scenario is called *chaotic inflation* (Linde, 1983). It is so simple that it is very difficult to understand why it was not discovered 20 years ago. The reason may have been purely psychological. We were hypnotized by the glorious past of the Big Bang theory. We were assuming that the whole Universe was created at the same moment, that initially it was hot, and that the scalar field from the very beginning was at the minimum of its potential energy density. Once we began relaxing these assumptions, we immediately found that inflation is not an exotic phenomenon invoked by theorists for solving their problems. It is a very general regime that occurs in a wide class of theories including many models with polynomial and exponential potentials $V(\phi)$.

Quantum Fluctuations as the Origin of Structure Formation in the Universe

Solving many difficult cosmological problems simultaneously by a rapid stretching of the Universe may seem too good to be true. Indeed, if all inhomogeneities were stretched away, what about galaxies? The answer is that while removing previously existing inhomogeneities, inflation at the same time created new ones. According to quantum field theory, empty space is not entirely empty. It is filled with quantum fluctuations of all types of physical fields. The fluctuations can be regarded as waves of physical fields with all possible wavelengths. If the values of the fields, averaged over a macroscopically large time, vanish, the space filled with the fields seems to us empty and can be called a vacuum.

In the inflationary universe the vacuum structure is more complicated. Waves that have very short wavelengths do not know that the Universe is curved; they move in all directions with a speed approaching the speed of light. However, inflation very rapidly stretches these waves. Once their wavelengths become sufficiently large, the waves begin feeling that the Universe is curved. At this moment they stop moving because of the effective viscosity of the expanding Universe with respect to the scalar field.

The first quantum fluctuations to freeze are those with large wavelengths. The amplitude of the frozen waves does not change later, but their wavelengths grow exponentially. In the course of the expansion of the Universe, new and newer fluctuations become stretched and freeze on top of each other. At that stage, we cannot call the waves quantum fluctuations. Most of them have exponentially large wavelengths. The waves do not move and do not disappear when averaged over large periods of time. What we get is an inhomogeneous distribution of the classic scalar field ϕ that does not oscillate. Only later, after the end of inflation, the waves, one after another, begin moving again. These inhomogeneities are responsible for perturbations of density in our Universe and for the subsequent appearance of galaxies (Mukhanov and Chibisov, 1981; Hawking, 1982; Starobinsky, 1982; Guth and Pi, 1982; Bardeen, Steinhardt, and Turner, 1983). Thus, inflation takes sub-subatomic quantum fluctuations and blows them up to the sizes of future galaxies and clusters (see Chapter 1).

The theory resembles the legend about the famous traveler Baron Munchausen (Raspe, 1988). Once he was hunting in winter. At night it became so cold that he could not blow his hunting horn — the sound was frozen inside it. In the morning when the horn became warm again, the sound was "defrosted," and suddenly the horn started playing all by itself. If inflationary theory is correct, galaxies have been created by the sound of "heavenly music" produced and frozen during inflation.

Experimental Tests of Inflationary Theory

In addition to explaining many features of our world, inflationary theory makes several important predictions. First, most versions of inflationary theory predict that the Universe should be extremely flat. However, several versions of inflationary theory exist that incorporate an open or a closed universe as well (Linde and Mezhlumian, 1995). Also, the density perturbations (as well as **gravitational waves**) generated during inflation have a very peculiar property — waves with wavelengths in the interval from 10^{10} to 10^{11} cm give approximately the same contribution to density perturbations as the waves with wavelengths from 10^{11} to 10^{12} cm, or from 10^{27} to 10^{28} cm, and so on. These specific predictions remove many uncertainties that plagued the theory of galaxy formation and provide several ways to test the inflationary theory.

Flatness of the Universe can be experimentally verified because the density of a flat Universe is related in a very simple way to the speed of its expansion given by the Hubble constant. So far observational data are consistent with this prediction. Properties of density perturbations produced during inflation can be tested in ways other than by looking at the distribution of matter in the Universe. The perturbations make their imprint on microwave background radiation. They make the temperature of this radiation slightly different in different places in the sky. This is exactly what was found 3 years ago by the Cosmic Background Explorer (COBE), and later confirmed by several other groups of observers (see Chapter 1).

It is too early to claim that that the results of COBE have confirmed inflationary theory, but it is certainly true that the results already at their level of accuracy could definitively disprove most of the inflationary models, and it did not happen. At present, no other theory can simultaneously explain why the Universe is so homogeneous *and* predict the small inhomogeneities ("ripples in space") discovered by COBE.

Nevertheless, we should certainly continue the search for alternative theories to explain the wonders of our Universe. Moreover, inflationary theory itself is changing with the rapid development of elementary particle theory. Many different inflationary models have been suggested, some of which are based on investigation of more complicated theories with several different scalar fields. The list of new models includes "extended inflation," "natural inflation," "hybrid inflation," and the like. Each model has its own unique features that can be experimentally tested, but most models are based on the same physical principles as the simple chaotic inflation model discussed above.

Self-Reproducing Inflationary Universe

The most recent development of inflationary cosmology is the theory of an eternally existing, self-reproducing, inflationary universe. This theory is rather general, but it looks especially surprising and leads to dramatic consequences in the context of the chaotic inflation scenario (Linde, 1986). We can visualize quantum fluctuations of the scalar field in the inflationary universe as waves, which first move in all possible directions, and then freeze on top of each other. Each freezing wave slightly increases the value of the scalar field in some parts of the Universe, and slightly decreases this field in other parts of the Universe.

Let us consider the rare parts of the Universe in which the freezing waves always increase the value of the scalar field ϕ, persistently pushing the scalar field uphill, to greater values of its potential energy $V(\phi)$. This is a very strange and obviously improbable regime. Indeed, the probability that the field ϕ will make a jump up (instead of down), is equal to one-half; the probability that the next time it will also jump up is also one-half; thus, the probability that the field ϕ without any special reason will make N consecutive jumps in the same direction is extremely small; it will be proportional to $1/2^N$.

Normally, we neglect such fluctuations. However, in this case they can be extremely important. Indeed, those rare domains of the Universe in which the field jumps high enough begin expanding exponentially with ever-increasing speed. The inflationary universe expands as e^{Ht}, where the Hubble constant is proportional to the square root of the energy density of the field ϕ. In our simple model with $V(\phi)$ proportional to ϕ^2, the Hubble constant H will be simply proportional to ϕ. Thus, the higher the field ϕ jumps, the faster the Universe expands. Very soon the rare domains in which the field ϕ persistently climbs the wall acquire a much greater volume than the domains that keep sliding to the minimum of $V(\phi)$ in accordance with the laws of classical physics.

From this theory it follows that if the Universe contains at least one inflationary domain of a sufficiently large size, it begins unceasingly producing new inflationary domains. Inflation in each particular point may cease very quickly, but there will be many other places that continue expanding exponentially. The total volume of all inflationary domains will grow without end. We can visualize this process by considering each inflationary domain as an exponentially rapidly inflating balloon. After a while, most parts of the balloon stop inflating, but in a few places in the balloon the Universe expands much faster than it was expanding before, and new inflationary balloons appear. These parts of the Universe in turn produce other inflationary balloons, and this process goes on as a chain reaction. This process is called *eternal inflation*. In this scenario, the Universe looks not like a single expanding ball, but like a tree of balls unceasingly producing new balls (see Color Plate 7.3).

In this scenario the Universe as a whole is immortal. Each particular part of the Universe may appear from a singularity somewhere in the past, and it may end up in a singularity somewhere in the future. However, there is no end for the evolution of the whole Universe. The situation with the beginning is less certain. It is most probable that each part of an inflationary universe originated from some singularity in the past. However, at present, we do not have any proof that all parts of the Universe were created simultaneously in a general cosmological singularity, before which there was no space and time. Moreover, the total number of inflationary balloons on our cosmic tree grows exponentially in time. Therefore, most balloons (including our own part of the Universe) grow indefinitely far away from the root of the tree. This removes the possible beginning of the whole Universe to the indefinite past.

We have made an attempt to illustrate this process by computer simulations of the evolution of the scalar field in an inflationary universe (see Color Plate 7.4). A series of images shows the time evolution of a scalar field ϕ in an inflationary universe. We begin with an almost homogeneous domain of space filled by the scalar field ϕ, which is shown as height of the golden surface in a two-dimensional universe. The Universe expands, but we shrink the image to the original size to show how the field evolves. At each step of the calculations, the value of the scalar field slightly decreases, but it also experiences quantum jumps that occasionally put it higher in some parts of the Universe. That is why we see the value of the scalar field decrease in most parts of the Universe, in accordance with the laws of classical physics, but there exist some places in which quantum fluctuations make this field large again. In the places that correspond to the peaks of the mountains on our figures, the Universe expands with extremely great speed, and eventually there is a lot of new space in which inflation can continue. We live in one of the valleys in which the scalar field rolled down. We see ourselves surrounded by a homogeneous part of the Universe, with small ripples of density and temperature, as recently discovered by COBE. However, far away from us there are places in which the field ϕ is still jumping high and producing new parts of the inflationary universe. To show that the Universe expands with a greater speed in the places in which the scalar field is large, it would be necessary to enlarge the size of each of the peaks in Color Plate 7.4. This would give us a picture similar to that in Color Plate 7.3, in which each of the peaks is represented by an exponentially large and rapidly expanding bubble.

We have considered the simplest inflationary model, with only one scalar field ϕ. In realistic models of elementary particles there are many different scalar fields. For example, in the unified theories of weak and electromagnetic interactions at least two other scalar fields exist, Φ and H. In some versions of these theories, the potential energy density of the fields has about a dozen different minima of the same depth. During inflation, these fields, just as the field ϕ, jump in all possible directions because of quantum fluctuations. After inflation, they fall to different minima of their energy density in different exponentially large parts of the Universe. Scalar fields change properties of elementary particles and the laws of their interaction. This means that after inflation the Universe becomes divided into exponentially large domains with different laws of low-energy physics. We have illustrated this point in Color Plate 7.3 by representing domains of the Universe with different laws of low-energy physics as exponentially large balloons of different colors.

Computer simulations show how the Universe becomes divided into different regions with different laws of physics inside each region. For example, consider a theory in which the field Φ, which is responsible for symmetry breaking in the theory, has a potential energy with three different minima. We will show the field in the first minimum by the color red, in the second minimum by green, and in the third minimum by blue. Quantum fluctuations of the field Φ are represented by the fluctuations of the color. The field φ, which is responsible for inflation, is shown, as in Color Plate 7.4, by the height of the "mountains." We begin our simulations with a part of the Universe in which the field Φ sits in one of the minima of its potential energy (red), and the field φ is homogeneous (we have a flat distribution of this field) (see Color Plate 7.5). After a few calculations we obtain mountains of the field φ divided into regions with all possible colors. Finally, in Color Plate 7.6, we obtain many sharp peaks, each of which can be interpreted as a new Big Bang (or, to be more precise, as a beginning of a new branch of inflationary domains in our cosmic tree).

Note that in the valleys, where the field φ is small, all fluctuations are not very strong. Therefore, in such regions, the scalar field Φ cannot jump from one minimum of its energy density to another. In this case the new parts of an inflationary universe remember the genetic code of their parents; that is, the color in the valleys does not change. However, if fluctuations are sufficiently large, mutations occur, and the new mountains (i.e., new inflationary bubbles) change their color, which means that the laws of low-energy physics in the daughter universes may differ from the laws in the parent universes. In some inflationary models, quantum fluctuations become so strong that even the effective number of dimensions of space and time can change. According to these models, we find ourselves inside a four-dimensional domain with our kind of physical laws not because domains with different dimensionality and with different particle properties are impossible or improbable, but simply because our kind of life cannot exist in other domains. A typical size of each such domain is so huge that for practical purposes we talk about division of our Universe into many separate universes; cosmic travel from one such domain to another would take longer than 10 billion years.

What is the probability of finding in our Universe a domain with a given volume containing space with given properties such as density of matter, laws of physics, colors? The answer to this question, which we have obtained in the context of a very wide class of inflationary theories, appears to be quite surprising (Linde, Linde, and Mezhlumian, 1994). We have found that very soon after the beginning of inflation the Universe approaches a stationary regime in which the relative volume of domains with different properties does not depend on time. This result has a simple interpretation in terms of our cosmic tree consisting of many inflationary bubbles of all possible colors. If we take a slice of the tree at time t_1, we find many green, red, and blue bubbles. At time t_2, we see even more such bubbles, but the ratio of the volume of green bubbles to the volume of red or blue ones remains constant. Each bubble at this tree evolves in time. After the end of inflation in each of the bubbles, the interior evolves in accord with the standard Big Bang theory. However, the Big Bang theory does not describe the whole Universe. Some of the most important properties of the Universe appear to be time-independent.

BACK TO THE FUTURE

The future of the Universe in the standard Big Bang theory was rather bleak. We had only two choices: to die in the fire in a collapsing closed Universe or to die in the cosmic cold of empty flat or open space. If we ignore quantum effects, inflationary theory leaves us with no choice at all. According to this theory, our part of the Universe resembles a part of a flat Friedmann Universe. It will continue expanding for exponentially large time, until all stars become dim and all protons decay.

Is there any way around this tragic outcome? First, proton decay by itself does not make the situation entirely hopeless. We can contemplate the possibility of "positron recycling" — converting them back to protons. Such processes should occur if the gas of positrons is compressed to a density three orders of magnitude greater than nuclear density and uses monopoles as catalysts of baryon formation. In 10^{35} years from now this process should be technologically feasible. However, in the absence of new sources of energy, maintaining this process for an indefinitely long period of time will be impossible.

Another possibility is suggested by the unusual structure of an inflationary universe. Is it possible to fly from our part of the Universe, when it begins to die, to faraway parts of the Universe that continue to be created now? We do not know the final answer to this question, but preliminary investigation suggests that at the time we would reach the parts of the Universe being created now, they would be as old as the part of the Universe from which we started our journey.

One more possibility is to create a new universe in a laboratory, jump in, and escape the problems of our world. Some time ago this would have seemed absolutely impossible because the Universe we would want to create should be huge, but now we know that inflation can easily make small things extremely big. Simple estimates in the context of the chaotic inflation scenario suggest that less than 1 mg of matter may be needed to initiate an eternal process of the Universe's self-reproduction. However, the theory of this effect is not quite reliable; some reasons exist that make this process entirely impossible. And even if it were possible to bake new universes the way we bake diamonds, we will be unable to jump into a new universe because at the moment of its creation it is microscopically small and extremely dense; later it would likely entirely decouple from our space.

Thus, we have not found any simple way to avoid the death of our civilization. But we have learned something unexpected, something that allows us to look at the cosmic drama from a totally different perspective. The process of a self-reproducing inflationary universe is somewhat similar to what happens to all of us. Some time ago we were born. We will die, and the whole world of our thoughts, feelings, and memories will disappear. However, there were those who lived before us, and there will be those who will live after us, and humanity as a whole, if it is clever enough, may live for a very long time. Now we know that even if our civilization dies, there will be other places in the Universe in which life will appear again, in all its possible forms. This gives us a possibility to have a slightly more optimistic attitude toward the future of our Universe. And we still have 10^{35} years to think about it.

REFERENCES

Albrecht, A., and Steinhardt, P. J. 1982. Cosmology for grand unified theories with radiatively induced symmetry breaking. *Physical Review Letters 48*: 1220–1223.

Bardeen, J., Steinhardt, P. J., and Turner, M. 1983. Spontaneous creation of almost scale-free density perturbations in an inflationary universe. *Physical Review D28*: 679–692.

Guth, A. H. 1981. The inflationary universe: A possible solution of the horizon and flatness problems. *Physical Review D23*: 347–356.

Guth A. H., and Pi, S.-Y. 1982. Fluctuations in the new inflationary universe. *Physical Review Letters 49*: 1110–1113.

Guth, A. H., and Weinberg, E. 1983. Could the universe have recovered from a slow first order phase transition? *Nuclear Physics B212*: 321–364.

Hawking, S. W. 1982. The development of irregularities in a single bubble inflationary universe. *Physics Letters 115B*: 295–297.

Kirzhnits, D. A., and Linde, A. D. 1972. Macroscopic consequences of the Weinberg-Salam model. *Physics Letters 42B*: 471–475.

Kirzhnits, D. A., and Linde, A. D. 1976. Symmetry behavior in gauge theories. *Annals of Physics (NY) 101*: 195–238.

Linde, A. D. 1982. A new inflationary universe scenario: A possible solution of the horizon, flatness, homogeneity, isotropy and primordial monopole problems. *Physics Letters 108B*: 389–393.

Linde, A. D. 1983. Chaotic inflation. *Physics Letters 129B*: 177–181.

Linde, A. D. 1986. Eternally existing self-reproducing chaotic inflationary universe. *Physics Letters 175B*: 395–400.

Linde, A. D. 1990a. *Particle Physics and Inflationary Cosmology* (Chur, Switzerland: Harwood).

Linde, A. D. 1990b. *Inflation and Quantum Cosmology* (Boston: Academic Press).

Linde, A. D. 1994. The self-reproducing inflationary universe. *Scientific American 271(5)*: 48–55.

Linde, A. D., Linde, D. A., and Mezhlumian, A. 1994. From the Big Bang theory to the theory of a stationary universe. *Physical Review D49*: 1783–1826.

Linde, A. D., and Mezhlumian, A. 1995. Inflation with $\Omega \neq 1$. *Physical Review D52*: 6789–6804.

Mukhanov, V. F., and Chibisov, G. V. 1981. Quantum fluctuations and "nonsingular" universe. *JETP Letters 33*: 532–535.

Raspe, R. E. 1988. *The Travels and Surprising Adventures of Baron Munchausen* (Sawtry, UK: Dedalus).

Starobinsky, A. A. 1979. Relict gravitational radiation spectrum and initial state of the Universe. *JETP Letters 30*: 682–685.

Starobinsky, A. A. 1980. A new type of isotropic cosmological models without singularity. *Physics Letters 91B*: 99–102.

Starobinsky, A. A. 1982. Dynamics of phase transition in the new inflationary universe scenario and generation of perturbations. *Physics Letters 117B*: 175–241.

GLOSSARY

Abiotic synthesis. Production of organic molecules by nonbiological processes (i.e., chemical reactions in the absence of living systems).

Absorption line. A wavelength (or small range of wavelengths) at which the brightness of a spectrum is lower than it is at neighboring wavelengths.

Accretion. The transfer of matter to the surface of a star, often from a companion star in a binary system. When the transferred matter goes into orbit around a star or black hole, an accretion disk is formed.

Active galaxy. A galaxy whose nucleus emits large quantities of electromagnetic radiation that does not appear to be produced by stars.

Adjustable parameter. In an equation a constant that can take on a variety of numerical values (sometimes including zero) so that the equation yields a family of related solutions, not just one solution.

Ambipolar diffusion. Process in which magnetic fields slowly drift outwards through a plasma.

Amino acids. Nitrogen-containing acids, some of which make up the building blocks of proteins.

Ångstrom (Å). A unit of length commonly used for visible wavelengths of light; $1 \text{ Å} = 10^{-8}$ cm.

Angular momentum. A measure of the amount of spin of an object; dependent on the object's rotation rate, mass, and mass distribution.

Antiparticle. A particle whose charge (if not neutral) and certain other properties are opposite those of a corresponding particle of the same mass. An encounter between a particle and its antiparticle results in mutual annihilation and the production of high-energy photons.

Astronomical unit (AU). Average distance between Earth and the Sun (150,000,000 km).

Baryonic matter. Protons, neutrons, and electrons, the components of ordinary matter.

Big Bang. The birth of the Universe in a very hot, dense state about 10 to 20 billion years ago, followed by the expansion of space.

Binary pulsar. A pulsar in a binary system. Often this term is used for systems in which the pulsar's companion is another neutron star.

Binary X-ray sources. Binary stars in which a white dwarf, a neutron star, or a black hole accretes matter from a normal companion, which produces large quantities of X-rays.

Binding energy. The difference in mass, expressed in energy units, when two or more particles are closely bound to each other compared with when they are far apart.

Bipolar outflow. A strong, well-collimated outflow of material from a young stellar object. The outflow is narrow in angular extent and is generally confined to the rotational poles of the system.

Blackbody. An object with a constant temperature that absorbs all radiation that hits it.

Black hole. A region of space-time in which the gravitational field is so strong that nothing, not even light, can escape. Predicted by Einstein's general theory of relativity.

CCD. Charge-coupled device. A solid-state imaging chip whose properties include high sensitivity, large dynamic range, and linearity.

Centrifugal force. The outward force felt by an object in a rotating frame of reference.

Cepheid. A type of pulsating variable star with a luminosity that can be determined from the period of its variation. Cepheids with long pulsation periods are bigger and thus more luminous than short-period Cepheids.

Chandrasekhar mass. Maximum stable mass for a degenerate star (white dwarf) or degenerate iron core in a massive star. Configurations more massive than the Chandrasekhar mass — about 1.4 solar masses, depending on composition — collapse or explode. Also called Chandrasekhar limit.

Chemical evolution. 1) Gradual buildup of the heavy element supply in a galaxy and accompanied by changes in the star formation rate, color, and luminosity of the galaxy; 2) Complex of chemical reactions that must occur on the surface of Earth or other inhabited planet before self-replicating molecules and biological evolution can appear.

Chemical reactions. Reactions that assemble or destroy chemical compounds through lending or sharing of electrons among atoms.

Cherenkov radiation. Light produced by charged particles moving through a transparent medium with a speed exceeding the local speed of light.

Circumstellar disk. A nebula of gas and dust orbiting a star. The circumstellar disks produced by the star formation process typically have radial sizes comparable to that of our solar system.

CNO tri-cycle. Linked cycles of nuclear reactions that transform hydrogen to helium, with carbon, nitrogen, and oxygen nuclei acting as catalysts.

Cold dark matter. A type of dark matter that was moving at much less than the speed of light 10,000 years after the Big Bang (see "Dark matter").

Column density. The total mass per unit area along a given line of sight through an astrophysical object such as a molecular cloud, a protostellar envelope, or an accretion disk. Proportional to the number of molecules (or dust grains) per unit area along a given line of sight.

Convection. A process that transports heat (energy) by motions of the fluid itself.

Core. In a main-sequence star, approximately the central 10%, by mass. In an evolved star, usually refers to the degenerate central region.

Core of molecular cloud. A small dense region within a molecular cloud. The cores are the actual birth sites of stars. The cores are much smaller than the cloud, but much larger than a star.

Cosmic microwave background. Diffuse glow of light, in the microwave (radio) part of the spectrum, left from the cooling Big Bang at an age of a few hundred thousand years.

Cosmic rays. Very high-energy protons and other particles found throughout interstellar space; their impacts on Earth's atmosphere make carbon-14 and cause mutations.

Cosmological constant. In Einstein's general relativity equations, a term that leads to an acceleration of the expansion of the Universe.

Cosmology. The study of the overall structure and evolution of the Universe.

Critical density. Density the Universe would have if its expansion rate were just barely sufficient to prevent a recollapse.

Dark matter. Apparently unseen matter that dominates the mass of the Universe. Many astronomers believe dark matter is made of a gas of exotic, weakly interacting particles. According to this theory, embryonic fluctuations in the matter density of the particles grew into galaxies, galaxy clusters, and galaxy superclusters.

Degenerate gas. Gas that is so dense its electrons can no longer move freely through space; the gas pressure depends only on the gas density, not on its temperature.

Density. Mass divided by volume.

Density fluctuation. A region in which the amount of matter per unit volume is slightly higher or lower than average. Somewhat analogous to a high or low pressure zone in Earth's atmosphere.

Dipole. A pattern where one side of the sky is hot and the opposite side is cold.

Dipole field. The pattern of electric field lines produced by a pair of equal and opposite electric charges, or of magnetic field lines surrounding a bar magnet.

Deoxyribonucleic acid (DNA). Nucleic acid with the structure of a double helix that serves as the genetic memory for life.

Doppler. Nineteenth century physicist who discovered the variation in the wavelength of waves caused by the motion of the source of the waves relative to the observer.

Doppler shift. The change in wavelength or frequency produced when a source of waves and the observer move relative to each other. Blueshifts (to shorter wavelengths) and redshifts (to longer wavelengths) are associated with approach and recession, respectively.

$E = mc^2$. Einstein's famous formula for the equivalence of energy and mass.

Eccentricity. A measure of the deviation of an elliptical orbit from a circle.

Electromagnetic force. One of the four forces of nature. Electromagnetic interactions hold electrons in atoms, hold atoms in molecules, and are used in all electronic devices.

Electromagnetic radiation. Self-propagating, oscillating, electric and magnetic fields; from shortest to longest wavelengths, the categories include gamma rays, X-rays, ultraviolet, optical (visible), infrared, and radio.

Electron. Negatively charged, low-mass particle that orbits atomic nuclei in atoms; normally the number of electrons is the same as the number of protons.

Electron degeneracy. A peculiar state of matter at high densities in which, according to the laws of quantum mechanics, electrons move very rapidly in well-defined energy levels and exert tremendous pressure on their surroundings.

Electron volt (eV). Energy necessary to raise an electron through a potential difference of one volt. One watt for 1 second is equivalent to 6 billion-billion eVs. Using the conversion $E = mc^2$, the rest mass of an electron is about 511,000 eV.

Electroweak. A unified force that combines the electromagnetic and weak nuclear interactions. Predicted by Weinberg and Salam and experimentally verified by Rubbia and van der Meer.

Elliptical galaxy. A system of 1 billion to 100 billion stars bound together by gravity. Ellipticals have basically round shapes, but typically have one long axis and one short axis, which may differ by a factor of 2; their true shapes are either like onions, footballs, or both. Unlike spiral galaxies, they have no thin disks of stars, gas, and dust and appear to have formed few new stars in the recent past. The stars in an elliptical have a wide variety of orbits in various directions, which leads to an overall appearance of random motion.

Emission line. A wavelength (or small range of wavelengths) at which the brightness of a spectrum is higher than it is at neighboring wavelengths.

Erg. A unit of energy (work) in the metric system equal to a force of one dyne (g-cm/s^2) acting through a distance of 1 cm; 10^7 ergs = 1 Joule.

Escape velocity. The minimum speed which an object must have to escape the gravitational pull of another object.

Event horizon. Boundary of a black hole from within which nothing can escape (*see* Schwarzchild radius).

Expansion of the Universe. The observed characteristic of the Universe that galaxies are all receding from each other with speeds that are in direct proportion to their separations. Thus, a galaxy twice as far from Earth as another galaxy will have a recession velocity twice as great as the closer galaxy. Galaxies do not actually move through space; space itself is expanding and thus carrying galaxies along like sequins painted on an expanding balloon. Also called Hubble expansion.

Fossil record. Entire collection of traces and remains of past life on Earth.

Fractal. A pattern having the mathematical property that any small part of it, viewed at any magnification, shares the same statistical character as the original, and is thus indistinguishable from the whole. Such patterns are common in nature (e.g., a coastline).

Fusion. Nuclear reaction in which two or more light nuclei combine to form a heavier one; some fusion reactions in stars are also called nuclear burning or simply burning.

Galaxy. A large, gravitationally bound collection of stars such as the Milky Way galaxy, which consists of several hundred billion stars. Galaxy shapes are generally spiral or elliptical, and sometimes irregular or peculiar.

Gamma ray. A very high-energy photon, more energetic than an X-ray.

Gas chromatograph mass spectrometer. An instrument that separates gases in a mixture and then determines the molecular mass of each component so that the component can be identified.

G dwarf problem. Near the Sun, the unexpected paucity of long-lived stars with less than about 10% of the solar metallicity.

General theory of relativity. Albert Einstein's comprehensive theory of mass, space, and time. According to the theory, the gravitational field associated with matter produces a curvature of space-time in its vicinity.

Grand unified theory (GUT). A model for unifying the strong nuclear force, the weak nuclear force, and the electromagnetic force into a single interaction. Several GUTs have been proposed, but not experimentally verified.

Grating. A piece of glass having many parallel grooves cut into its surface. Used to disperse light into its component wavelengths.

Gravitational waves. According to relativity theory, waves ("ripples" in space-time) emitted because of changes in the distribution of matter.

Great Attractor. A proposed nearby supercluster of galaxies in a region of the sky largely obscured by the Milky Way, and thus difficult to study.

Hα. Photon produced by a hydrogen atom when its electron jumps down from the third to the second energy level.

Habitable planets. Planets with liquid water, a supply of elements needed for biochemistry, and an environment stable for billions of years.

Half-life. The time it takes for half a given quantity of a radioactive substance to decay.

Heavy bombardment. Rain of asteroidal and comet-like material (**planetesimals**) that comprised the final formation of the planets.

Homogeneous. The same at all locations (e.g., homogenized milk is not separated into cream and milk).

Horizon. Edge of the *visible* Universe, but not the actual edge of the Universe (because the Universe has no edge).

Hot Big Bang. A model of the Universe beginning at very high density and temperature, which expands and cools to become like the Universe we observe now.

Hot dark matter. A type of dark matter that was moving at a substantial fraction of the speed of light 10,000 years after the Big Bang (see "Dark matter").

Hydrothermal vents. Fissures, usually on the ocean floor, through which hot water and reactive gases emanate.

Inflationary scenario. A modification of the Big Bang model in which a large cosmological constant exists temporarily early in the history of the Big Bang and leads to a rapid accelerating expansion of the Universe, which is then followed by the normal Big Bang model with a decelerating expansion.

Initial mass function (IMF). Distribution of stellar masses of a given stellar population at its moment of birth; the IMF is not the same as the distribution of stellar masses seen later because stars with different masses have different lifetimes.

Interstellar medium. Space between the stars; filled to some extent with gas and dust.

Ionized. Having lost at least one electron. Atoms become ionized primarily by the absorption of energetic photons and by collisions with other particles.

Isothermal equation of state. A relation between the pressure P, the density ρ, and the temperature T of a gas. When the temperature of a gas remains the same, an isothermal equation of state has the form $P = a^2\rho$, where the sound speed a is a constant.

Isotopes. Atomic nuclei having the same number of protons (and, hence, almost identical chemical properties) but different numbers of neutrons and, therefore, somewhat different masses.

Isotropic. The same in all directions.

Kelvin (K). The size of 1 degree on the Kelvin temperature scale, in which absolute zero is 0 K, water freezes at 273 K, and water boils at 373 K. To convert from the Kelvin scale to the Celsius (centigrade; C) scale, subtract 273 from the Kelvin scale value.

Keplerian orbits. Orbits that obey the classical laws of Kepler.

Kepler's third law. If one object orbits another, the square of its period of revolution is proportional to the cube of the major axis of the elliptical orbit.

Large Magellanic Cloud (LMC). A dwarf companion galaxy of our Milky Way galaxy, located about 170,000 light years away; best seen from Earth's southern hemisphere.

Lichen. A life-form based on the symbiotic association of fungus and algae.

Light curve. A plot of an object's brightness as a function of time.

Light-year. Distance light travels through a vacuum in 1 year; about 10 trillion km.

Luminosity. Energy per unit of time generated by an astrophysical object.

Magnitude. A scale used by astronomers to measure flux. Each 5 units on the magnitude scale corresponds to a 100-fold decrease in the flux. The Sun has magnitude −26.5. Sirius, the brightest star in the night sky, has magnitude −1.6. The faintest stars visible with the naked eye have magnitude 6.

Main-sequence. The phase of stellar evolution, lasting about 90% of a star's life, during which the star fuses hydrogen to helium in its core.

Marine diatoms. A silicon-shelled unicellular organism found in ocean water.

Metallicity. The fraction by mass of a star, galaxy, or gas cloud that is made up of all elements heavier than helium (not just actual metals).

Microbial mats. Layers of microorganisms found on the bottom of lakes and tide pools.

Millisecond pulsar. A pulsar whose period is roughly in the range 1 to 20 milliseconds.

Molecular cloud. A large cloud of interstellar gas in the molecular state. These clouds form stars but are much larger than a single star; the clouds typically have masses between 10^4 and 10^6 $M_\odot$.

N-body (computer) simulation. Following the growth of structure through a computer program that calculates the gravitational force between N bodies (where N is a large number) representing the total mass.

Nebula. A region containing an above-average density of interstellar gas and dust.

Neutralino. A particle predicted by supersymmetric models for the forces of nature. The models predict that each type of known

particle will have a supersymmetric partner. The neutralino is the lightest, electrically neutral, supersymmetric partner and is a candidate for cold dark matter. As of 1995, no supersymmetric partner particles of any kind have been observed experimentally.

Neutrino. A massless, or nearly massless, uncharged particle that interacts exceedingly weakly with matter after being created by certain nuclear reactions. There are three types: electron, muon, and tau neutrinos.

Neutron. Uncharged, massive particle found in nuclei of atoms; different isotopes of a given element have different numbers of neutrons in their nuclei.

Neutron degeneracy. Similar to electron degeneracy, but with neutrons replacing electrons. Sets in at a much higher density and pressure than electron degeneracy.

Neutron star. The compact endpoint in stellar evolution in which 1 to 2 $M_\odot$ of material is compressed into a small (diameter = 20–30 km) sphere supported by neutron degeneracy pressure.

Nonbaryonic. Not made up of neutrons and protons, and thus not like any of the known chemical elements.

Nova. Nonfatal stellar explosion caused by burning of degenerate hydrogen on the surface of a white dwarf in a binary system.

Nuclear fusion. Reactions in which low-mass atomic nuclei combine to form a more massive nucleus.

Nuclear reaction. A reaction that transforms one type of atomic nucleus into another by changing the number of neutrons and protons.

Nucleosynthesis. The creation of elements through nuclear reactions.

Obliquity. The angle between the spin axis of a planet and the direction perpendicular to its orbital plane. The larger the obliquity, the stronger are seasonal climate variations.

Opacity. The degree to which matter restricts the flow of electromagnetic radiation; a perfectly transparent substance has zero opacity.

Organic material. Compounds and molecules that contain carbon.

Oxidants. Compounds that cause binding with oxygen.

Panspermia. The idea that life was carried to Earth from elsewhere.

Parsec. A unit of distance used by astronomers, equal to 3.085678×10^{13} km (3.26 light-years).

Phase transition. A change between two different phases of matter, such as a solid melting and becoming a liquid, or a liquid boiling to become a gas.

Photon. A quantum, or package, of electromagnetic radiation that travels at the speed of light; from highest to lowest energies the categories include gamma rays, X-rays, ultraviolet, optical (visible), infrared, and radio.

Photosynthesis. Production of organic material by sunlight.

Phylogenetic tree. Relationship by descent of a group of organisms.

Planet. An object of substantial size (larger than about 1000 km in diameter), but having a mass less than roughly 0.01 $M_\odot$, that orbits a star.

Planetary nebula. A shell of gas, expelled by a red giant near the end of its life (but before the white dwarf stage), which glows because it is ionized by ultraviolet radiation from the star's remaining core.

Planetesimals. See **Heavy bombardment**.

Positron. Antiparticle of an electron.

Pre-main-sequence star. A young star that is not generating energy exclusively through the fusion of hydrogen. Pre-main-sequence stars initially generate energy through gravitational contraction and eventually evolve into stellar configurations in which hydrogen fusion takes place.

Primordial black holes. Submicroscopic black holes that may have been produced during, or shortly after, the Big Bang.

Primordial soup. Mixture of abiotically produced molecules, thought to have led to the origin of life.

Progenitor. In the case of a supernova, the star that will eventually explode.

Proteins. A class of biomolecules constructed from sequences of amino acids.

Proton. Positively charged, massive particle in nuclei of atoms; different chemical elements have different numbers of protons in their nuclei.

Proton-proton chain. Sequence of nuclear reactions that transforms hydrogen to helium.

Protostar. A star that is still in the process of forming. The star itself is generally deeply embedded within an infalling envelope of dust and gas.

Pulsar. An astronomical object that is detected through pulses of radiation (usually radio waves) having a short, extremely well-defined period; now thought to be a rotating neutron star with a very strong magnetic field.

Quantum fluctuations. The uncertainty principle in quantum mechanics leads to all allowed interactions occurring with some probability (see **vacuum energy density**).

Quantum mechanics. A twentieth-century theory that successfully describes the behavior of matter on very small scales (such as atoms) and radiation.

Quasar. The nucleus of an active galaxy. This nucleus is unusually luminous, perhaps 10 to 1000 times more powerful than the rest of the galaxy. Their ultimate energy source is suspected to be accretion onto a massive black hole. Most quasars are at distances of billions of light-years from Earth.

Radioactive nucleus. A nucleus capable of spontaneously emitting an electron, a helium nucleus, or a gamma ray.

Radiogenic heating. Heating caused by energy released by the decay of radioactive elements.

Red giant. Evolutionary phase following the main-sequence of a solar-type star; the star becomes both large and bright (though cool on the surface). Hydrogen burns in a shell around a helium core.

Redshift. Doppler shift for objects receding from Earth causes the wavelengths of light to get longer, and hence visible light shifts toward the red part of the spectrum. Lengthening of waves can also be caused by propagation in a strong gravitational field.

Reducing atmosphere. An atmosphere that is rich in molecules that contain hydrogen, for example, ammonia (NH_3) and methane (CH_4).

Relative humidity. Ratio of actual vapor pressure of water to its saturation vapor pressure at a given atmospheric pressure and temperature.

Rest mass. The mass of an object that is at rest with respect to the observer. Massless particles, such as photons, *must* travel at the speed of light, but particles having nonzero rest mass *cannot* reach the speed of light.

Ribonucleic acid (RNA). Information-transporting molecule related to DNA; involved in protein synthesis.

Schwarzschild radius. 1) Radius to which a given mass must be compressed to form a nonrotating black hole. 2) Radius of the event horizon of a nonrotating black hole.

Sedimentary deposits. Material accumulated by water or wind.

Shock wave. A compressional wave, characterized by a discontinuous change in pressure, produced by an object traveling through a medium faster than the local speed of sound.

Singularity. A mathematical point of zero volume associated with infinite values for physical parameters such as density.

Solar mass ($M_\odot$). The mass of the Sun, 1.99×10^{33} g, about 330,000 times the mass of Earth.

Space-time. The four-dimensional fabric of the Universe whose points are events having specific locations in space (three dimensions) and time (one dimension).

Spallation. Breaking up or eroding of something fairly solid; in particular, the breakup of heavy atomic nuclei when hit by cosmic rays.

Spectral energy distribution. A graph showing how much energy (in radiation) is emitted by an astronomical object as a function of the frequency (or wavelength) of the light.

Spectrum. A plot of the brightness of electromagnetic radiation from an object as a function of wavelength or frequency.

Spiral galaxy. A galaxy, like our Milky Way, made up of 1 billion to 100 billion stars bound together by gravity. The spiral pattern, found in the thin disks of stars, gas, and dust that surround a spheroidal bulge of stars, is highlighted by sites of continuing star birth. The stars in the bulge swarm in many directions, as in the closely related elliptical galaxies; stars within the disk trace near-circular orbits around the center of the galaxy.

Steady State. A model of the expanding Universe with constant density and all other physical properties. Because of the expansion of the Universe, matter must be continually created to maintain constant density.

Stellar population. Mix of stars of different masses, with their various temperatures, metal abundances, and ages, that makes up a large system such as a star cluster or galaxy.

Stellar wind. Continuous or quasicontinuous release of gas from the outer atmosphere of a star.

Strong nuclear force. One of the four forces of nature. The strong nuclear force holds the particles in the nucleus of atoms together.

Supergiant. Evolutionary phase following the main-sequence of a massive star; the star becomes very bright and cool (red) or hot (blue); a sequence of nuclear reactions occurs in the stellar interior.

Supernova. Violent explosion of a star at the end of its life. Hydrogen is present or absent in the spectra of Type II or Type I supernovae, respectively.

Supernova remnant. Cloud of chemically enriched gases ejected into space by a supernova.

Tidal force. Difference between the gravitational force exerted by one body on the near and far sides of another body.

Time dilation. According to relativity theory, the slowing of time perceived by an observer watching another object moving rapidly or located in a strong gravitational field.

T Tauri star. A newly formed (pre-main-sequence) star with many characteristic signatures of youth, including strong emission lines and excess infrared radiation. The name derives from the star T Tauri, which is the prototype for this class of very young stars.

Ultraviolet light. Light with wavelengths shorter than blue light, that is, wavelengths between 200 and 300 nm.

Vacuum energy density. Quantum theory requires empty space to be filled with particles and antiparticles being continually created and annihilated. This could lead to a net density of the vacuum, which if present, would behave like a cosmological constant (see **quantum fluctuations**).

Variable star. A star whose apparent brightness changes with time.

Virgo Supercluster. A large concentration of galaxies in the direction of the constellation Virgo, with a recession velocity of 10^3 km/s with respect to the Milky Way.

Weak nuclear force. One of the four forces of nature. The weak nuclear force is responsible for some radioactive decays as well as some of the fusion reactions in the Sun that provide heat and light for Earth.

White dwarf. Evolutionary end point of stars that have initial masses less than about 8 times the solar mass. All that remains is an electron degenerate core of helium, carbon-oxygen (the majority of cases), or oxygen-neon-magnesium.

Wormhole. Connection between two black holes in separate universes or in different parts of our Universe. Also called Einstein-Rosen bridge.

X-ray nova. A nova that emits an unusually large amount of X-ray radiation.

INDEX

A

Abiotic synthesis, 113
Absorption lines, 69
Abt, H. A., 49
Accretion, 72
Accumulation of planetesimals, planet formation by, 54, 55–56
Active disks, 47, 48
Active galaxies, 85
Adams, Fred C., 37, 38, 39, 44, 45, 48, 50, 51, 59
Age of Universe, and Hubble constant, 12
Albrecht, Andreas, 134
Alpher, R. A., 8
Aluminum, 98, 101–102
Ambipolar diffusion, 38
    core formation through, 40–41
Amino acids, 109–110
Ammonia, 114, 121, 123
Andromeda nebula, 6, 16, 26
Angular momentum, 21, 38, 50, 79
    transport of, 53, 54
Antarctica, 119–120
Antielectrons, 76
Antineutrinos, 73, 76, 80
Antiparticles, 73
Appenzellar, I., 47
Approximations, 105
    closed box, 104
    homogeneous, 103, 104
    one zone, 103
Aquarius, 66
Archaea, 110
Argon, 89
Arons, J., 40
ASCA (X-ray satellite), 85
Asteroids, 55, 68, 102, 121
Astronomical units (AU), 121

B

B^2FH, 94, 99, 100, 101
Baade, Walter, 22, 73, 77
Bacteria, 110
Balbus, S. A., 53
Bardeen, J., 135
Baryonic matter, 23
Beckwith, S., 48
Bell (now Burnell), Jocelyn, 77, 81
Benz, W., 51
Beryllium, 89, 101
Bethe, Hans, 94, 96
Biemann, K., 113
Big Bang theory, 1, 13, 15, 132, 134, 138
    and black holes, 84, 86
    and complexity, 29–33
    and dark matter, 11–12
    and expansion of Universe, 2, 3, 5, 16
    and future of Universe, 128–129
    and galaxy formation, 21–26
    and horizons, 9
    and hydrogen and helium, 101
    and inflation, 9–10
    vs. inflationary cosmology, 127–128
    and infrared background, 12
    and light element abundances, 7–8
    nucleosynthesis, 90–93
    problems of, 129–130
    and stellar explosions, 67
"Big Crunch," 5
Binary pulsars, 81
Binary X-ray sources, 84
Binding energy, 73
Biology, development of, 33
Bipolar outflow phase of evolution, 38
Blackbody, 6
Black holes, 61, 67, 81–83
    detecting, 84–85
    fun facts about, 83–84
    myths about, 85–86
    primordial, 84
Blitz, L., 39, 58
Bodenheimer, P., 51, 52
Boron, 101
Brahe, Tycho, his supernova of 1572, 72, 74
Brown dwarfs, 57
Bulge, of galaxy, 18, 22
Burbidge, E. M., 94
Butcher, Harvey, 26
Butner, H. M., 44

C

Cairns-Smith, A. G., 114
Calcium, 89, 104
Calvet, N., 44, 48
Cameron, A. G. W., 100
Carbon, 89, 90, 94, 95, 100, 105
    burning, 97–98
    and CNO tri-cycle, 96
    and helium burning, 97
    isotopes of, 112
    and life on Earth, 123
Cassen, P., 42
Cassiopeia A, 74
Center for the Study of Evolution and the Origin of Life (CSEOL) symposium, 15
Centrifugal forces, 78
Cepheid variables, 4, 16, 78
Chadwick, James, 73
Chandrasekhar, S., 98
Chandrasekhar mass limit, 72, 98
Chaos, 23
Chaotic inflation, 132–134, 136
Charge-coupled device (CCD), 68–69
Chemical evolution, galactic, 102–107
Chemical reactions, 89
Chemistry, development of, 33
Cherenkov radiation, 76
Chibisov, G. V., 135
Chlorine, 89
Chyba, C., 121
Circumstellar disks, 38, 47–48, 49
    gravitational instabilities in, 50–51
    physics of, 49–54
    planet formation within, 54–57
    viscous evolution of, 53
Climate stability, 121
Clow, G. D., 119
CNO tri-cycle, 96, 101
Cold dark matter, 12
Column density, 44–45

Comets, 55, 115, 120, 121
Complexity, 28–34
Computers, and galaxy formation, 23, 24, 26–27
Convection, 47, 77
Copi, C., 8
Core, of molecular clouds, 38
    formation of, through ambipolar diffusion, 40–41
    initial conditions for collapse of, 42
Core, of star, 65–66
COsmic Background Explorer (COBE) satellite, 6, 7, 10–11, 12, 13, 135, 137
"Cosmic burps," 66
Cosmic microwave background radiation, 5–7, 23
Cosmic rays, 101
Cosmological constant, 4, 10
Cosmological principle of homogeneous Universe, 2, 130
Cosmology, 67
Couch, Warrick, 26
Crab nebula, 79–80
Critchfield, C. L., 94
Critical density, 3, 128
Cygnus X-1, 84–85

D

Dark matter, 5, 11–12, 23, 106
    cold, 12
    hot, 11, 12
    "hot and cold running," 12
    nonbaryonic, 11
Daugherty, D. A., 39
Davis, W. L., 114, 115, 117, 120
Degenerate gases, 97
Density, 66
    fluctuations in, 24
    vacuum energy, 10
DesMarais, D. J., 115
Deuterium, 8, 47, 90, 93, 96, 101
Dipole, 6
Dipole field, 79
Disk accretion, 46, 49–50, 54
DNA, 110, 115
Doppler shift, 2, 16, 69
Dressler, Alan, 15
    Voyage to the Great Attractor, 34
Dust grains, 55–56, 57

E

Earth, 20, 38, 48, 55, 66
    and accumulation of planetesimals, 56
    changes in obliquity on, 121
    life on, 109–116, 118, 122–123
Eccentric orbit, 119
Eigen, M., 115
Einstein, Albert, 4, 10, 65, 131, 132
    general theory of relativity of, 78, 82, 128
Einstein-Rosen bridge, 86
Electromagnetic (nuclear) interaction, 10, 130–131
Electromagnetic radiation, 67
Electron degeneracy, 66
Electrons, 90
Electron volts (eV), 7, 76–77
Electroweak theory, 10
Elementary particles, theories of, 129, 130–131, 134, 136
Element synthesis, in stars, 93–94, 101–102
    and heavy elements burning, 97–98
    and helium burning, 96–97
    and hydrogen burning, 94–96
    and iron, 99–100
Elliptical galaxies, 18, 21, 22, 26–27, 71
Ellis, Richard, 26
Emerson, J. P., 48
Emission lines, 69
Encrenaz, T., 113
Entropy, 33–34
    war between gravity and, 37, 60–61
Ergs, 76
Escape velocity, 4, 82
Eternal inflation, 136
Eukarya, 110
Europa, 120
eV. *See* Electron volts
Evans, Robert, 68
Event horizon, 82
Evolution, defined, 15
Expansion of the Universe, 1–5, 16
Extrasolar neutrino astrophysics, 76

F

Feynman, Richard, 89
Filippenko, Alexei V., 65
Fluctuations in density, 24
Fluorine, 89, 98, 101
Forveille, T., 50
Fossil record, 110
Fowler, Willy, 65n, 72
Fractal, 130
Freedman, W., 4
Friedmann, Alexander, 128, 138
Friedmann, E. I., 119
Fuller, G. A., 42, 48
Fusion, 95

G

Galactic halo, 106
Galaxies
    active, 85
    defined, 16–20
    elliptical, 18, 21, 22, 26–27, 71
    formation of, 20–22
    irregular, 18, 21, 72
    lenticular, 18
    making real calculations of formation of, 23–26
    S0 (S-zero), 18
    spiral, 18–20, 21, 22, 26–27, 71, 72
    supernovae vs., 67
    Type I supernovae in, 71
Galaxy harassment, 27
Galileo Galilei, 16
Gamma ray, 7
Gamow, G., 8, 90, 100
Garcia, Francisco, 68
Gas chromatograph mass spectrometer (GCMS), 117
G dwarf(s), 104, 105, 106
    problem, 104, 105
General theory of relativity, 78, 82, 128
Giant planets, 55
Gold, 89
Goodman, A. A., 39
Grand unified theory (GUT), 10, 76
    supersymmetric (Susy GUT), 12
Grating, 69
Gravitational instabilities, 50–51, 54
    planet formation by, 54–55, 56–57
Gravitational waves, 78
Gravity, 33
    and formation of galaxies, 20–21
    war between entropy and, 37, 60–61
Great Attractor, 6, 34
GS 2000+25, 85
Guth, A., 9, 133–134, 135

H

Habitable planets, 107
Haldane, J. B. S., 112–113
Half-life, 75
Hart, M. H., 122
Hartmann, L., 44, 48
Hashisu, I., 51
Hawking, S., 84, 135
Hawley, J. F., 53
Hayashi, C., 53

Index

"Heat death," 33
Heavy bombardment, 110
Heavy elements, 55, 57, 67
    burning, 97–98
Hebes Chasma, 120
Heiles, C. H., 39
Helium, 8, 89, 90, 93, 94, 101
    burning, 95, 96–97, 98, 100
Helix nebula, 66
Herman, R., 8
Hertzsprung-Russell (HR) diagram, 44, 46, 47
Heterotrophy, 114
Hewish, Antony, 77, 81
Homogeneous Universe, 2, 130
Horizons, 8–9
"Hot and cold running dark matter," 12
Hot Big Bang model, inflationary scenario in, 1, 9–11. *See also* Big Bang theory
Hot dark matter, 11, 12
Houlahan, P., 39
Hoyle, F., 53
Hubble, Edwin, 1, 4, 16–18
Hubble constant, 12, 132, 135, 136
    age of the Universe and, 12
    discovery of, 98
    and expansion of Universe, 1–5
Hubble Space Telescope (HST), 4, 26–27, 80, 85
Hulse, Russell, 81
Hydrogen, 89, 90, 93, 101
    burning, 94–96, 98, 100
Hydrothermal vents, 116

I

Inflation, 9–11
    chaotic, 132–133, 134, 136
    eternal, 136
Inflationary cosmology, 1, 9–11, 24
    Big Bang theory vs., 127–128
    brief history of, 133–134
    experimental tests of, 135–136
    and quantum fluctuations as origin of structure formation in Universe, 134–135
    and self-reproducing inflationary Universe, 136–138, 139
    simplest version of, 132–133
    and unified theories of elementary particles, 130–131
Infrared background, 12–13
Initial mass function (IMF), 37, 57
    constant, 103–104
    observed, 57–58
    scaling relation for, 59–60
    as two-phase process, 58–59
    variable, 106
Instantaneous recycling, 103, 104, 105
International Astronomical Union, Central Bureau for Astronomical Telegrams of, 69
International Ultraviolet Explorer, 74
Interstellar medium, 66
Ionized gas, 66
Iron, 89, 94, 95, 97, 99–100, 104
Irregular galaxies, 18, 21, 72
Irvine, W. M., 113, 120
Irvine-Michigan-Brookhaven (IMB) collaboration, 75–76
Isothermal equation of state, 51
Isotopes, 75, 90, 112
Isotropic Universe, 2

J

Jovian planets, 55
Joyce, G., 115
Jupiter, 55, 120, 121, 122

K

Kamiokande, 75–76
Kant, Immanuel, 16, 54
Kastner, J. H., 50
Katz, Neal, 27
Kenyon, S., 44, 48
Kepler, Johannes, 67, 74
    third law of, 78
Keplerian disk rotation curve, 50
Kirzhnits, D. A., 134
Kissel, J., 113
Knacke, R. F., 113
Knoll, A. H., 116
Krueger, F. R., 113
Kuiper Airborne Observatory, 74
Kushner, D., 116

L

Lada, C. J., 38, 44, 46, 48
Ladd, E. F., 44
Lake, George, 27
Lake, J. A., 110
Landau, Lev, 73
Landau Institute, 133
Laplace, P. S. de, 54, 82
Large Magellanic Cloud (LMC), 67, 74–75, 104
Las Campanas Observatory, 74
Laskar, J., 121
Laughlin, G. P., 51, 52
$L\text{-}A_V$ diagram for protostars, 44–46
Lazcano, A., 113–114
Lebedev Institute, 134
Lehninger, A. L., 110
Lenticular galaxies, 18
Leuschner Observatory, 69
Levy, E. H., 53
Lichens, 116
Light curve, 69
Light element abundances, 7–8
Light-years, 67
Lin, D. N. C., 53, 54
Linde, A. D., 9, 127, 128, 134, 135, 136, 138
Linde, D. A., 138
Lithium, 93, 101
Lizano, S., 38, 39, 41, 42, 59
Luminosity, 44, 45
    protostellar, 38, 43
Lundmark, K., 98
Lüst, R., 53
Lynden-Bell, D., 53

M

M51, 69
M81, 68
M83, 68
M87, 85
Magnesium, 89, 101
Magnetic fields, 39, 41, 53–54
Magnitude, 2, 70
Maher, K. A., 121
Main-sequence, 44, 65, 101
Manganese, 89
Marine diatoms, 120
Mars, 55, 73, 110, 122, 123
    absence of organics on, 113
    Antarctica and, 119–120
    life on, 109, 117–120
Max, C., 40
McCaughrean, M. J., 57
MCG-6-30-15, 85
McKay, C. P., 109, 114, 115, 116, 117, 119, 120, 122
Melosh, H. J., 114
Mercury, 55, 56, 66, 110
Merrill, P., 100
Metal-enhanced star formation (MESF), 106
Metallicity, 105, 106
Meteorites, 102, 115
Methane, 114, 121
Mezhlumian, A., 138

Michell, John, 82
Microbial mats, 110
Microwave background, 5–7, 23
Milky Way galaxy, 1–2, 6, 18, 26, 105
    habitable planets in, 107
    and infrared background, 12, 13
    number of stars in, 20, 102
    and pulsars, 77
    size of, 21
    and supernovae, 68, 73–74
    and white dwarfs, 72
Miller, G. E., 57
Miller, S. L., 113–114
Millisecond pulsars, 81
Minimum mass solar nebula, 55
Minkowski, Rudolph, 70
Models
    with bells and whistles, 104–106
    simple, 103–104
Molecular clouds, 38–49, 59–60
    cores of, 38
    defined, 38
    initial conditions for collapse of cores of, 42
    and spectrum of initial conditions, 58
    star formation in, 38–49
Monopoles, magnetic, 129
Moon, 110, 121
Moore, Ben, 27
Moosman, A., 42
Mount Wilson Observatory, 16, 22, 70
Mouschovias, T., 39, 40
Mukhanov, V. F., 135
Mundt, R., 47
Myers, P. C., 39, 42, 44

N

Nakano, T., 41
N-body computer, 24, 26–27
Nebula, 65
    planetary, 66, 97
Nebular hypothesis, 54
Neon, 89, 98, 105
Neptune, 55, 56, 121
Neutralino, 12
Neutrinos, 11, 12, 73, 75–77
    three types of, 7, 8, 11
Neutron degeneracy, 77
Neutrons, 90
Neutron stars, 61, 67, 75–76, 77–79, 86
    and millisecond and binary pulsars, 81
    observational evidence of, 79–80
Newton, Sir Isaac, 4, 23, 24, 82, 85
NGC 4258, 85

NGC 4526, 69
Nitrogen, 89, 97, 100, 105
    and CNO tri-cycle, 96
Nobel Prize, 81, 89, 94, 98
Noble gases, 89
Nonbaryonic dark matter, 11
Nonlinear simulations of star/disk systems, 51–52
Nova, 67, 101
Nuclear fission, 65
Nuclear reactions, 66, 90
Nucleosynthesis, 72, 90
    Big Bang, 90–93

O

Obliquity, 121
Oemler, Gus, 26
Olympus Mons, 117
Omnicentric Big Bang model, 2
Opacity, 75
Oparin, A. I., 112–113
Oppenheimer, J. R., 77
Organic material, 113
Orion nebula, 16, 68, 69
Oró, J., 115
Owen, T., 122
Oxidants, 117
Oxygen, 89, 94, 95, 100, 105, 122
    buildup of, on Earth, 115–116
    burning, 98
    and CNO tri-cycle, 96
    and helium burning, 97

P

Palla, F., 46
Palmer, P., 39
Palomar Observatory, 70
Panspermia, 114
Papaloizou, J. C., 53, 54
Parsec, 4
Passive disks, 47–48
Peebles, Jim, 21
Penzias, A. A., 5, 128
Phase transition, 10
Phosphorus, 89, 98
Photons, 74, 130–131
Photosynthesis, 110–112
Phylogenetic tree, 110
Pi, S.-Y., 135
Planck density, 127, 130
Planetary nebula, 66, 97
Planetesimals, 55, 110
    planet formation by accumulation of, 54, 55–56

Planet formation
    by accumulation of planetesimals, 54, 55–56
    by gravitational instability, 54–55, 56–57
Planets, 66
    giant or Jovian, 55
    habitable, 107
    terrestrial, 55
Pluto, 55
Plutonium, 99
Population I stars, 106
Population II stars, 106
Positrons, 76
Potassium, 89
Pre-main-sequence phase of evolution, 38
Pre-main-sequence stars, 47–48
Primordial black holes, 84
Primordial soup, 114
Pringle, J. E., 53
Progenitor stars, 70
Prokaryotes, 110
Proteins, 110
Proton-proton (p-p) chain, 96
Protons, 90
Protostars, 42, 43
    L-A_V diagram for, 44–46
Protostellar collapse, 42–43
Protostellar phase of evolution, 38
Protostellar radiation, 43–44
Protostellar to stellar transition, 46–47
PSR 1937+21, 81
PSR 1953+29, 81
Pulsars, 77–80
    binary, 81
    millisecond, 81

Q

Quantum-mechanical pressure, 66
Quasars, 85

R

Radiation
    Cherenkov, 76
    electromagnetic, 67
    protostellar, 43–44
Radioactive nuclei, 72
Radiogenic heating, 120
Raspe, R. E., 135
Red giants, 66, 96
Redshift, 2, 3, 83–84
Reducing gases, 113
Relative humidity, 116
Rest mass, 76

Reynolds, R. T., 120
Rivera, M. C., 110
RNA, 110, 115
Rucinski, S. M., 47
Ruden, S. P., 50, 53
Russell, D. A., 116
Rydgren, A. E., 47
Ryle, Sir Martin, 81

S

S0 (S-zero) galaxies, 18
Sagan, Carl, 68, 113, 120, 121
Sagittarius, 18
Saha, A., 4
Salpeter, E. E., 57, 59, 120
Sanduleak-69° 202, 74
Saturn, 50, 54, 55, 120
Scalar fields, theory of, 131, 134
Scalo, J., 39, 57
Schidlowski, M., 112
Schneider, J., 122
Schopf, J. W., 110
Schramm, D., 8
Schwarzschild, Karl, 82
Schwarzschild radius, 82
Search for Extraterrestrial Intelligence (SETI) program, 122
Sedimentary deposits, 112
Séguin, R., 116
Selenium, 89
Self-reproducing inflationary Universe, theory of, 136–138, 139
Semi-empirical mass formula, 58–59
Sharples, Ray, 26
Shelton, Ian, 74
Shock waves, 67
Shu, F. H., 38, 39, 41, 42, 43, 44, 46, 47, 48, 50, 59
Silicon, 98, 123
Silk, J., 59
Silver, 89
Singularity, 82–83
  cosmological, 127, 129
Sky & Telescope, 74
Sleep, N. H., 121
Smail, Ian, 26
Small Magellanic Cloud, 74, 104
Smooth particle hydrodynamics (SPH) computer code, 51
SN 1987A, 67, 73–77, 80
SN 1993J, 68
SN 1994D, 69
SN 1994I, 69
Sodium, 89, 98, 101, 104
Solar Maximum Mission satellite, 75

Solar system, life in, 120–122
Spallation, 101
Spectral energy distributions, 44, 49
Spectrum of initial conditions, 58
Spiral galaxies, 18–20, 21, 22, 26–27, 71, 72
Spiral patterns, with one spiral arm, 50–51
Spitzer, L., 39
Stahler, S. W., 42, 43, 46, 47
Star/disk systems, 50
  nonlinear simulations of, 51–52
Star formation, 104, 105–106. *See also* Initial mass function
  in molecular clouds, 38–49
Starobinsky, A. A., 9, 133, 135
Stars, searching for life around other, 122–123. *See also* Element synthesis in stars
Steady State model of the Universe, 4–5, 6, 9, 10
Steinhardt, P. J., 9, 134, 135
Stellar birthline, 47
Stellar explosions, 67–68
Stellar populations, 22, 106–107
Stellar transition, protostellar to, 46–47
Stellar winds, 66
Stepinsky, T. F., 53
Stevenson, D. J., 121
Stoker, C. R., 117
Strong nuclear force, 10
Strong nuclear interactions, 130–131
Sulfur, 89
Sun, 20, 38, 43, 48, 49, 73
  death of, 66
  elements contained in, 55
  energy generated by, 65
  and initial mass function, 57
  mass residing in, 50
Supergiants, 73, 74–75, 96
Supernova(e), 67, 86
  classification, 69–73
  searching for, 68–69
  SN 1987A, 67, 73–77, 80
  Type I, 69, 70, 71–72, 73, 77
  Type II, 69, 70, 72–73, 74, 77
Supernova remnants, 68
Supersymmetric grand unified theories (Susy GUTs), 12
Surface temperature, 44

T

Taam, R. E., 42, 43
Tarantula nebula, 74
Taylor, Joseph H., 81

Technetium, 100
Terebey, S., 42
Terrestrial planets, 55
Thorium, 89, 100
Tidal forces, 83
Time dilation, 83
Tinsley, Beatrice M., 102–103, 104, 106
Titan, 113, 120
Titanium, 104
Tohline, J. E., 51
Tremaine, S., 50
Trimble, Virginia, 89, 94
T Tauri phase of evolution, 38, 46, 50
T Tauri stars, 47, 49
Turbulence, 39, 59–60
Turner, M., 8, 135
Tycho. *See* Brahe, Tycho

U

Uranium, 89, 99, 100
Uranus, 55

V

V404 Cygni, 85
Vacuum energy density, 10
Valles Marineris canyon system, 120
Variable stars, 74
Vela nebula, 79, 80
Venus, 55, 123
Viking spacecraft, 117
Virgo cluster, 6, 69, 85
Viscous accretion, 50
  disks, evolution of, 53
Visual extinction, 45–46
Volkoff, G., 77

W

Walker, C. K., 44
Walker, G. A. H., 121
Wallis, M. K., 120
Water, 89, 115, 116, 117, 120
  vapor, atmospheric, 122–123
Weakly Interacting Massive Particle (WIMP), 11–12
Weak nuclear interactions, 10, 130–131
Weather forecasting, 103
Weinberg, Eric, 134
Weinberg-Salam model, 10
Wetherill, G. W., 121
Whirlpool galaxy, 69

White dwarfs, 61, 66, 71–72, 77, 97, 98
Wilking, B. A., 57
Wilson, R. W., 5, 128
Wiseman, J. J., 39
Woese, C. R., 110
Wood, D. O. S., 39
Woodward, J. W., 51, 52

Wormhole, 86
Wright, Edward L., 1

X

X-ray nova, 85
X-ray sources, binary, 84

Z

Zak, D. S., 47
Zent, A. P., 117
Zinnecker, H., 57
Zuckerman, B., 39, 50, 121
Zwicky, Fritz, 70, 73, 77